HOW TO IDENTIFY MUSHROOMS TO GENUS V:
Cultural and Developmental Features

HOW TO IDENTIFY MUSHROOMS TO GENUS V: Cultural and Developmental Features

BY ROY WATLING

Roy Watling
Royal Botanic Gardens
Inverleith Row
Edinburgh EH3 LR5
Scotland

MAD RIVER PRESS INC.

© Roy Watling

Published by: Mad River Press, Inc.
                Rt. 2 Box 151 B
                Eureka, CA 95501

Printed by: Eureka Printing Co., Inc.
                106 T Street
                Eureka, CA 95501

ISBN 0-916422-17-8

PREFACE

Cultural techniques have only comparatively recently been introduced as an aid in the classification of agarics and boleti. Although far from an absolute requirement for identification, cultural studies are essential to enable the mycologist to understand more easily not only the limits of the genera of Agaricales but also their relationship with other groups of larger fungi. Cultural characters expand the range of ammunition available to the agaricologist wishing to unravel the relationships between different species of gill-fungi and their allies.

It is therefore considered that the time is ripe to produce an inexpensive book in which cultural techniques are documented and information such as description of asexual stages, poorly known before, discussed. It has been prepared with the idea of providing a source of information with which the student's field studies can be expanded and blended and which allows them to take on a new meaning. It is in this way that it is hoped to encourage a better understanding of the complete mushroom. The book has been written as a companion to the other volumes in this series, allowing the student to develop a command of the complete field of agaricology. It is hoped that the text has been arranged in such a way as to allow it to be of use also to teachers wishing to study agarics in the laboratory and so be useable as a laboratory manual.

My thanks are offered to my wife Elizabeth for her help during all stages of the preparation of the manuscript, and for the production of the final typescript.

This book is dedicated to my assistants at the Royal Botanic Garden, Edinburgh, both past and present, who at one time or another carried out most if not all the techniques outlined between these covers, to Ms. Celia Alden, Heather McGilvary, Janis Sweeney, and Norma Gregory. Through their hard work the techniques have been tested and any small problems ironed out and refinements added.

TABLE OF CONTENTS

Preface i

I. INTRODUCTION 1

II. RECIPES 3
   A. Culture Media 3
   B. Nutrient and Buffer solutions 11
   C. Antibacterial Supplements 13
   D. Micro-stains and Slide mountants 14
   E. Sectioning Recipes 17
   F. Preservatives 20

III. TECHNIQUES 21
   A. Field Observations 21
      1. COLLECTING TECHNIQUES 21
      2. PRESERVATION OF VOUCHER MATERIAL 22
         a. Temperate countries 22
         b. Sub-tropical - tropical areas 25
   B. LABORATORY 26
      1. ROUTINE LABORATORY TECHNIQUES 30
      2. TECHNIQUES FOR SECURING PURE CULTURES . . . 30
         a. Isolates from basidiome (fruit-body) tissue 31
            (i) Isolation from hymenium and pileus (cap) . . . 31
            (ii) Isolation from stipe (stem) and velar fragments . 32
         b. Isolation from vegetative phase 33
            (i) Isolation from soil 33
            (ii) Isolation from wood-samples 34
            (iii) Isolation from mycorrhiza 34
         c. Isolation from spores 35
            (i) Basidiospores 35
            (ii) Asexual spores 38
         d. Indirect methods 39
            (i) Soil dilution plates 39
            (ii) Soil plates 39
            (iii) Flotation method 39
            (iv) Hyphal isolation 40
      3. TECHNIQUES FOR GROWING AGARICS IN PURE CULTURE . . . 40
         a. Inoculating 40
            (i) Vegetative tissue 40
            (ii) Basidiome tissue 41
            (iii) Basidiospores and Mitospores . . 41
         b. Cultivation and Maintenance 41
            (i) Procedure after isolation 41
            (ii) Restricions and purification . . . 42

    4. Techniques for Maintenance of Cultures. 44
    5. Techniques for Inducing Fructification. 46
  C. Staining Techniques. .51
    1. Procedure. 52
    2. Alternative schedules. 54
    3. Schedule for staining nuclei 55
  D. Genetic Techniques 56
    1. Single-spore isolation 56
    2. Maceration 57
    3. De-dicaryotization 58

IV. CHARACTERISTICS OF AGARICS IN CULTURE 59
  A. Introduction. 59
  B. Development of basidiome. 59
    1. RANGE OF DEVELOPMENTAL TYPES 59
    2. EXAMPLES 63
  C. Secondary Spores 65
    1. RANGE OF MITOSPORE MORPHOLOGY 65
      a. Solitary holoblastic conidia. 66
      b. Thallic arthroconidia. 66
      c. Chlamydospores. 68
      d. Modified clamp-connections. 70
      e. Examples. .70
  D. Anatomical and morphological features of the vegetative hyphae. .72
    1. CATEGORIES OF FEATURES 72
      a. Bulbiloid bodies. 72
      b. Sclerotia and related structures 73
      c. Rhizomorphs and similar structures 74
      d. Sterile fruiting. 74
      e. Additional cultural characters. 74
      f. Bioluminescence 79
      g. Examples. .79
    2. CULTURE FORMULAE 81
      a. Criteria. .81
      b. Examples. .82
      c. Procedure .83

V. EXPERIMENTS, STUDIES AND TESTS 85
  A. Chemical Tests. 85
    1. GALLIC ACID TESTS 85
    2. BIOCHEMICAL TESTS 86
      a. Instant tests 86
      b. Short-term incubation tests 87
      c. Long-term incubation tests. 88
    3. ADDITIONAL REACTIONS—SPOT TESTS 89
    4. USE OF BASIDIOME MATERIAL 90
  B. Genetic Studies 90

B. GENETIC STUDIES continued
    1. INTRODUCTORY DISCUSSION. 90
    2. MATING OF KNOWN AND UNKNOWN STRAINS. 93
    3. MATING-TYPE DETERMINATION 94
    4. 'OIDIAL' HOMING; ITS USE IN CULTURAL STUDIES 95
    5. COMPARATIVE STUDIES. 96
    6. PHYSIOLOGICAL EXPERIMENTS 96

VI. SUGGESTED EXERCISES. 98
  A. LABORATORY EXPERIMENTATION. 98
    1. *Agaricus hortensis* & *A. brunnescens*. 98
    2. *Coprinus cinereus* 101
    3. *Stropharia ferrii* 102
    4. *Psilocybe cubensis* 103
    5. *Pholiota nameko* 103
    6. *Galerina mutabilis*. 103
    7. *Agrocybe cylindracea* 104
    8. *Volvariella esculenta*. 104
    9. *Flammulina velutipes* 105
    10. *Pleurotus ostreatus* and allies 106
    11. *Lentinula edodes*. 107
    12. *Schizophyllum commune*. 107
  B. ECOLOGICAL EXERCISE - COPROPHILOUS FUNGI 108

VII. APPENDICES
  A. TABLE OF GASTROMYCETE-AGARIC LINKS 111
  B. TABLE OF CYPHELLA-AGARIC LINKS 114
  C. GLOSSARY 116
  D. REFERENCES AND ADDITIONAL READING. 123
  E. CHECK LIST OF CHEMICALS AND EQUIPMENT 131
  F. LIST OF ABBREVIATIONS. 133

VIII. INDEX 134

IX. PLATES AND FIGURES 140

I. INTRODUCTION

Laboratory studies now contribute a very important part in the routine identification of micro-fungi and plant pathogens. Recently such studies have also become a standard technique in the study of the wood-rotting fungi, but this new technology has rarely been employed in studies of the fleshy mushrooms and toadstools. By growing these organisms under comparable artificial conditions, an assessment can be made as to the role played by environmental factors in determining the morphological characters so important in the identification of a species. Armed with such techniques an additional and powerful tool is now available to the agaric-taxonomist.

Growing organisms under artificial conditions is called <u>culturing</u>. In order to understand the cultural features of agarics and boleti, a review of the terminology is necessary. The terms for the macroscopic and microscopic features are taken from the other publications in this series, although in some cases they are explained again for clarity.

All agarics are classified in the Basidiomycotina. Like other basidiomycetes they are characterised by the production of sexual spores (<u>basidiospores</u>) which are borne on basidia. The basidium of the agaric is an undivided single cell, which generally produces four spores at its apex. It is known as a <u>holobasidium</u>. If the basidium is divided into four cells each producing a spore, the compound structure is termed a <u>phragmobasidium</u>. The possession of this last type of basidium is at present known neither in the agarics nor the boleti. The agarics forcibly disperse their basidiospores and cast a spore-print if the fruiting body is laid gills-down on a piece of paper or on a glass slide. However, some close relatives of agarics, usually classified in the gasteromycetes, do not discharge their spores forcibly. The basidium in these fungi is called an <u>apobasidium</u>.

When a basidiospore produced by the basidium germinates, it may either directly bud off asexual spores (<u>mitospores</u> or <u>conidia</u>) or produce a germ-tube which ultimately forms a long cylindrical tube or filament called a <u>hypha</u>. The production of a germ-tube is characteristic of the agarics and their allies. The hypha produced from the spore elongates and branches to form an interwoven mat of numerous filaments, which is termed the <u>mycelium</u>. It is this mycelium and its accompanying appendages which one studies in culture. The hyphae making up the mycelium frequently produce asexual propagules which either germinate to form other hyphae and ultimately a mycelium or fuse with already existing hyphae.

In the majority of agarics two haploid nuclei derived from different parents exist side by side in each cell constituting that organism. In the agarics these two nuclei are often paired for very long periods of time, and only in one cell, the basidium, do the two nuclei fuse and divide to fulfill the sexual process. This phenomenon is unique to the Basidiomycotina.

When the spores are allowed to germinate on prepared food material under controlled conditions, the result is a culture. The phenomenon parellels the cultivation of plants in pots except that in the laboratory all but the organism under study is eliminated, i.e., it is called an _axenic_ culture. Only infrequently are fungi grown in culture with other organisms, usually trees, bacteria or yeasts; even under these conditions the cultures are very carefully controlled. Cultures are also produced by growing hyphae extracted from the vegetative thallus of the fungus or from the reproductive body. Cultures produced from the same species but from different sources are called _isolates_. Differences between such isolates may indicate whether one is dealing with either different strains or varieties of that species; it may even be found that one is dealing with different species, or more excitingly, 'species' in their formation.

This book is arranged so as to facilitate the study of cultural and allied features with ease. The first part of the book covers the recipes for reagents, media and similar essential materials used in the techniques described, and includes the methods of preparation of the food material and basic laboratory technology.

The second part of the book describes how to prepare cultures of agarics and how to preserve voucher material, whilst the third part indicates the characteristics exhibited when agarics are in culture and those which can be demonstrated by utilising the techniques outlined earlier. The third part introduces the majority of readers to a whole new realm of features which will allow a mushroom in the field to be viewed in quite a different light from then on. These features include developmental characteristics and the production and variation found in the asexual spores which the agarics often form. This same part leads logically into a more specialised part covering chemical tests and simple genetic studies.

The concluding part of the book is limited to agarics which readily carry out their complete life-cycle under artificial conditions and so might act as a source for those wishing to carry out laboratory experimentation.

The appendices contrast markedly with the main parts, as might be expected. The first two offer tables which link the experimental data with field observations on agarics and other groups of basidiomycetes. The glossary provided will allow unfamiliar terms to be learned and added to the vocabulary, as well as acting as a link between the present publication and earlier parts in the series. When a particular subject takes one's fancy, then reference to the reading list will allow one to find a publication which will satisfy the curiosity; although the list is not exhaustive it will allow the readers to expand their knowledge. The final appendix is intended for those lacking laboratory facilities and who would like to study at home a few of the facets of these interesting organisms.

II. RECIPES
A. Culture Media

MEDIA CONTAINING AGAR

All recipes are made up with 1000 ml water, preferably distilled.

Natural media

Media containing 20-25 g agar-agar

1. Corn-meal agar, abbreviated CMA : see pg. 48.

   Ground maize   25 g

2. Malt-extract agar, abbreviated MA : see pgs. 37, 42, 48, 83.

   Malt extract   20 g

The basic media can be modified by:

2a. Addition of 10 g, 20 g, 50 g, or 100 g glucose (dextrose) or sucrose to give a range of media with various percentages of sugar : see pg. 45.

3. Peptone/dextrose/malt agar : see pg. 42.

Peptone	1 g
Malt extract	20 g
Dextrose	20 g

By replacing the dextrose by 0.006 g of o-phenyl phenol, the agar can be used for isolating fungi from woody tissue. The phenol is added from a stock solution prepared by dissolving 1 g of o-phenyl phenol in 99ml of 95% ethanol; see pg.

Media containing 15-20 g agar-agar

4. Malt-extract/dextrose agar, abbreviated MDA : see pg. 42.

Hydrated magnesium sulphate (Epsom salts)	5 g
Malt extract	5 g
Potassium hydrogen phosphate	5 g
Dextrose	5 g
Ammonium chloride	5 g
Ferric citrate	0.5 ml of a saturated aqueous solution
Peptone	1 g added if required

5. Potato/carrot agar, abbreviated PCA: see pgs. 42, 44, 49.

Scrubbed and grated potatoes	20 g
Peeled and grated carrots	20 g

6. Potato/dextrose agar, abbreviated PDA: see pg. 42.

Scrubbed and grated potatoes	200 g
Glucose (dextrose)	20 g

In recipes 5 & 6, commercial powdered dehydrated potato may be substituted for the fresh vegetable. Care must be taken only to use potato powder without added preservatives. Use at 6 g per 10 g of fresh material.

7. Gallic-acid agar: this *agar* is used in the study of wood-rotting fungi: see pg. 83.

Powdered malt extract	15 g
Gallic acid	5 g

Tannic acid may be substituted for the gallic acid if so desired.

8. Yeast-extract agar, abbreviated YA: see pg. 42.

Yeast extract	3 g
Hydrated magnesium sulphate	0.2 g
Dipotassium hydrogen phosphate	2 g
Glucose (dextrose)	10 g

The basic medium can be modified by the replacement of the sulphate and phosphate by 3 g each of malt extract and peptone to make Yeast/malt-extract agar.

9. Dung-extract agar or Lange's agar; based on a medium designed by C. H. Kauffman and utilised by members of the Michigan University staff: see pgs. 9, 48.

Dung extract	100 ml For preparation see pg. 9.
Maltose	5.0 g
Hydrated magnesium sulphate (Epsom salts)	0.5 g
Calcium nitrate	0.5 g
Peptone (or casein hydrolysate)	0.1 g
Dipotassium hydrogen phosphate	0.25 g

Useful agars are made by adding the same quantity of dung extract to PDA and PCA and to Oat agar (see below).

Media with 12-15 g agar-agar

10. Oat-meal agar: see pg. 42.

    Patent brand of porridge oats,
        oat-flakes etc.   40 g

11. Modess' modification of Hagem's medium; this agar is used in the study of mycorrhizal fungi: see pg. 34.

Malt extract	5 g
Glucose (dextrose)	5 g
Potassium dihydrogen phosphate	0.5 g
Hydrated magnesium sulphate (Epsom salts)	0.5 g
Ammonium chloride	0.5 g
Ferric (iron) chloride	10 drops of a 1% aqueous

solution prepared by dissolving 1 g of ferric chloride in 100 ml water. The addition of thiamine to give a final concentration of 25 ppm is recommended: see pg. 34.

12. Vegetable agar: see pg. 46.

Concentrated vegetable juice	180 ml
Calcium carbonate	2 mg
Distilled water	920 ml

Tomato juice or patent vegetable juice such as V-8 can be substituted for the vegetable juice.

Natural products incorporated into agar

    Any vegetable and/or fruit can be introduced into an agar medium. Vegetables or fruit should be soaked in distilled water for one or two hours before cutting up or grating prior to weighing.

The quantities suggested are:

Seed corn: cooked wheat, barley, or rice	80 g
Carrots	500 g
Beans	400 g
Peas	400 g
Beet	500 g
Celery	600 g
Prunes	25 g

See pgs. 35 & 46.

13. Bread agar; this agar is used in the study of bioluminescence; see pg. 79.

Commercial bread-crumbs	100 g
Agar-agar	18 g
Distilled water	1000 ml

The bread used in this recipe should not be one containing preservatives; a white or brown loaf from the local baker is admirable if allowed to dry and then 'crumbed'.

Compounded media for specialised work

14. Basidiomycete agar with 12 g agar-agar; this medium is used in spot tests: see pgs. 75, 86.

Malt extract	10 g
Potato/dextrose	100 ml of mixture prepared above (pg. 4) before adding the agar-agar.
Benomyl	5 ppm
Neomycin sulphate	50 ppm
Streptomycin sulphate	50 ppm

20 PDA commercially produced agar tablets can replace the 100 ml of PDA mixture if so desired.

Chemically defined media

15. Czapek Dox agar: see pg. 9.

Hydrated ferrous sulphate	0.01 g
Sodium nitrate	2 g
Potassium dihydrogen phosphate	1 g
Potassium chloride	0.5 g
Hydrated magnesium sulphate	0.5 g
Sucrose	30 g
Agar-agar	15 g

This medium can be modified by the addition of 1 ml of a solution of 5% aqueous yeast extract or a 4% aqueous malt extract per 1000 ml.
Steep agar is prepared by adding 1% corn-steep liquor to the dissolved salts.
Glucose (dextrose) may be substituted for the sucrose to advantage.

16. Fries' medium; this medium has been used by P. Day for genetic experiments with Coprinus: see pg. 101.

L-asparagine	2 g
Ammonium tartrate	1.5 g
Potassium dihydrogen phosphate	1 g
Disodium hydrogen phosphate	2.25 g
Sodium sulphate	0.29 g
Thiamine	40 g
Glucose (dextrose)	20 g

The basic medium is made up with 20 g of agar-agar and 1000 ml of water; it can be supplemented with:

- 16a. 0.75 g yeast extract
- 16b. 0.75 g bacteriological peptone
- 16c. 0.66 g malt extract

Antibiotic media made up with 10 g agar-agar

17. Rose bengal/streptomycin agar: see pg. 43.

Peptone	2 g
Potassium dihydrogen sulphate	0.5 g
Hydrated magnesium sulphate	0.5 g
Glucose (dextrose)	10 g
Rose bengal	50 g

Make up the medium in 1000 ml of water and add just before use 80 ml of a solution of streptomycin prepared by adding 1 g to 100 ml of water. Aureomycin may be substituted for the streptomycin if so desired.

18. Neopeptone/dextrose/rose bengal/aureomycin agar: see pg. 43.

Neopeptone	5 g
Glucose (dextrose)	10 g
Rose bengal	3.5 ml of an aqueous solution

Keep the powdered Rose bengal in the refrigerator and make up when required as 1 g in sterile water (100 ml); this solution is added to the medium before autoclaving. Aureomycin is added as a solution at 0.05 ml per 10 ml of agar medium after pouring. This solution is prepared by adding 1 g of aureomycin hydrochloride to 15 ml distilled sterile water; then 3.5 ml of this solution is added to 1000 ml before use. The stock solution may be kept in the refrigerator for up to 3 months.

Non-nutrient agar (Tap-water agar) & modifications to this preparation; see pg. 88.

19. Tap-water agar:

Agar-agar	12 g
Water	1000 ml

Before this medium solidifies in the Petri dish one can add sterile fragments of filter paper, vegetable plugs, lupin stems, wheat straws, three or four rabbit pellets, i.e. any natural product which will be a source of food, especially one chosen from ecological observations; see pg. 108.

Commercial media

Many of the above media may be purchased directly from various laboratory suppliers. They are delivered in tablet form, each tablet making 5 ml of medium; other quantities are made with tablets of other sizes. The medium is in a dehydrated state, and no agar-agar is required. One disadvantage of using these products is that the media are often deficient in nitrogen and only slow growth is obtained. This can be alleviated by adding 1.5 g of ammonium tartrate to every 1000 ml of medium prepared.

Additional media

For more estensive lists of recipes see Russell Stevens 'Mycology Guidebook', prepared by the Mycological Society of America; see refs.

Preparation of agar media

Place the chemicals, potato, carrot, or similar material in a flask in two-thirds the quantity of distilled water required by the recipe, and simmer over a water-bath or in a double-saucepan for one hour. Strain the solution through a sieve or through muslin cloth; it is preferable not to filter. It may be necessary to wring the cloth to force through all the material.

Dissolve the agar-agar in the rest of the water by adding a small quantity at a time. The dissolved agar-agar should then be added to the medium during further boiling. Finally add the sugars.

After cooking, the medium is strained and bottled. During bottling, the bulk of the medium should be retained in the water-bath and small quantities dispensed. About 15 ml of medium should be poured into 1 oz McCartney bottles or medical flats; the latter should be filled to within 25 mm of the shoulder. The bottles can be charged through a funnel which is set up in a clamped ring on a retort stand; the flow of the medium should be controlled by means of a Mohr clip over a short length of rubber tubing attached to the stem of the funnel.

After loosening the caps of the bottles, the medium can be sterilised either in an autoclave or a pressure-cooker. Failing the possession of either, sterilization can be achieved by using an enclosed steamer. With the first two methods sterilization should be conducted at 15 lb pressure per sq. inch for twenty minutes, but in the last it is necessary to boil for twenty minutes on each of three successive days. Leave the medium in the cooker until the pressure has fallen to normal, then take out and allow to cool in a clean area before screwing the caps down tightly.

A litre of medium will be sufficient to fill 60-65 '1 oz' McCartney bottles, 30 '2 oz' medical flats, or 15 '4 oz' flats.

If glass-distilled water (i.e. distilled using a glass still) is used to prepare the chemically defined media, then trace elements will undoubtedly be absent or deficient. This problem can be corrected by adding hydrated zinc sulphate at 0.01 g per 1000 ml and hydrated copper sulphate at 0.05 per 1000 ml or the trace element solution described below on pg. 11. Fortunately sufficient concentration of trace elements is found in natural products to eliminate this problem in the preparation of natural media. Czapek Dox agar is best buffered to pH 6.0 with a suitable buffer solution, of which some are outlined below; pgs.12, 13.

Whenever sucrose is used, always autoclave the chemical constituents separately from the sugar and then mix the ingredients, in order to avoid caramelization of sucrose and the production of unwanted precipitation.

Salts should be added to 300 ml of the 1000 ml and heated up with the agar-agar itself in 500 ml more of the water. Peptone and dextrose or similar compounds and mixtures of natural products are best added to the residual 200 ml and mixed with the other ingredients just before autoclaving.

Gallic acid agar is sometimes solidifies too slowly. To avoid this problem the gallic acid (5 g) should be added to 150 ml of hot, sterile water and incorporated into 850 ml of the malt agar whilst still warm.

The dung extract for Lange's medium is prepared by boiling one horse-dung 'apple' in 200 ml of distilled water in a flask for 1-2 mins. Filtering of the resultant liquid is not necessary unless single-sport isolation or hyphal-tip isolation is to be attempted. 100 ml of this extract is then added to 900 ml of water containing the salts and agar-agar indicated in the recipe: see pg. 56.

SOLID MEDIA USUALLY NOT INCORPORATING AGAR

20. Paper-pulp medium: see pg. 48-49.

Laminated paper-pulp should be soaked overnight in water, after which it is either mechanically or manually shredded. Sufficient of the paper-pulp so prepared is placed in the favoured glass container so as to fill it loosely to one third full. A nutrient agar of one's choice is then poured when molten over the pulp, the container closed by a cottonwool bung, and the whole autoclaved. The medium can be modified by adding a layer of sterile soil, peat mixture, etc. to the surface after vegetative growth is fully under way. Best results have been obtained by partially squeezing the paper-pulp pad free of most excess water before placing in the container.

21. Badcock's sawdust medium: see pgs. 48, 102.

Maize meal	50 g
Bone flour	30 g
Potato starch	17 g
Sucrose	2 g
Wood ash	1 g

The wood ash is prepared by combusting pine-wood, although beech, etc. can be used if available. Glucose can be substituted for the sucrose if so desired.

22. Etter's corn-meal medium: see pg. 48.

Corn meal	48 g
Corn starch	16 g
Powdered pine-wood	8 g

The medium may be modified by substituting wood of one's choice for the pine-wood. The medium should be moistened with a 2.5% solution of malt extract before autoclaving.

23. Corn-kernel medium: see pg. 49.

Maize kernels	50 g

Place maize into favoured container and cover with a 2% solution of malt extract before autoclaving.

24. Vermiculite/peat mixture medium: see pg. 47, 51.

Vermiculite	50 g
Peat moss	50 g

Inert coarse sand may be added to this medium if so desired. There will be a change in pH with autoclaving, the final value of which will depend on the constituents used. Sandy mixtures will become pH 6.5 $\geq$ 5.3, and mixtures with high vermiculite content will become pH 4 $\geq$ 5.2. The pH can be adjusted by application of a suitable buffer solution described below, pg. 12. Peatmoss is best sterilised by steaming for several days running, as outlined earlier on pg. 9 as opposed to autoclaving at 15 lb pressure for 20 min. The medium is used for growing plants along with agarics in sterile conditions and is preferably used with supplementary lighting at 700 foot-candles on a fourteen-hour photoperiodic pattern, if possible, and below 22°C. The mixture should have a final depth of about 5 cm in the base of the container.

B. Nutrient and Buffer Solutions

Trace-element solutions

1. Trace-element solution : see pg. 9, 11, 12.

Ferric chloride	200 mg
Copper sulphate	400 mg
Manganous chloride	140 mg
Zinc chloride	400 mg
Sodium molybdate	40 mg
Boric acid	60 mg

    The solution should be made up with glass-distilled water.

2. Useful quantities to know when making up media lacking trace elements

Manganese	for 0.5 ppm	2 g $MnSO_4$ per 100 ml.
Molybdenum	for 0.1 ppm	0.15 mg MoO_3 per 100 ml.
Zinc	for 0.05 ppm	0.23 g $ZnSO_4 \cdot 7H_2O$ per 100 ml.
Copper	for 0.02 ppm	0.089 g $CuSO_4 \cdot 5H_2O$ per 100 ml.
Boron	for 0.5 ppm	2.85 g HBO_3 per 1000 ml.

Salt solutions (See pg. 50)

3. Complete-salts solution

Sodium sulphate	4 g
Potassium dihydrogen phosphate	16 g
Potassium chloride	8 g
Hydrated magnesium sulphate	2 g
Calcium chloride	1 g
Trace elements as above #1	8 ml

The solution should be made up with 992 ml glass-distilled water.

4. Nutrient solution

Calcium chloride	0.05 g
Sodium chloride	0.025 g
Potassium dihydrogen phosphate	0.50 g
Diammonium hydrogen phosphate	0.25 g
Hydrated magnesium sulphate	0.15 g
Thiamine	25 µg
Glucose	2.5 g
Ferric chloride	1.2 ml of a 1% solution prepared by adding 1 g to 100 ml of water

    The nutrient solution should be made up to 1000 ml.

5. Knop's solution

Calcium nitrate	0.5 g
Potassium nitrate	0.125 g
Hydrated magnesium sulphate	0.125 g
Dipotassium hydrogen phosphate	0.125 g
Ferrous chloride	0.005 g

    Trace element solution (pg. 11) is optional, but for optimal results 8 ml should be added. The solution is made up to 1000 ml.

Buffer solutions : (See pgs. 9, 55).

6. Citric-acid buffer (suggested for use in chemical tests: see pg. 87)

    Mix 54.7 ml of a M/5 (1/5 molar) aqueous solution of disodium hydrogen phosphate (prepared by thoroughly mixing 8.89 g of powder with 1000 ml water) with 45.3 ml of a M/10 aqueous sol. of citric acid (prepared by mixing 21 g of citric acid in 1000 ml water). The resulting pH will be 5.4.

7. McIlvaine's standard buffer solutions

    The following mixtures of a) 0.1 molar solution of citric acid (prepared by adding 21.01 g to 1000 ml water) and b) 0.2 molar solution of disodium hydrogen phosphate (prepared by adding 35.6 g to 1000 ml water) are prepared for the pH indicated:

    pH 5.0   9.7(a) & 10.3(b)
    pH 5.5   8.62(a) & 11.38(b)
    pH 6.0   7.37(a) & 12.63(b)

pH 6.5   5.3(a) & 14.7(b)

Quantities are in ml.

8. Sørenson's glycine buffer

    25 ml of a 0.2 N solution of glycine (prepared by adding 11 g to 1000 ml water is added to 19.3 ml of a 0.2 N solution of sodium hudroxide (prepared by adding 8 g to 1000 ml water).
    The solution is made up to 100 ml with water and gives a pH of 10.4.

9. Sørenson's buffer mixture

    The following mixtures of a) disodium hydrogen phosphate (prepared by adding 11.9 g to 1000 ml water) and b) potassium dihydrogen phosphate (prepared by adding 9.1 g to 1000 ml water) are prepared for the pH indicated:

    pH 6.0   1.4(a) & 8.6(b)
    pH 6.5   3.5(a) & 6.5(b)
    pH 7.0   6.1(a) & 3.9(b)
    pH 7.5   8.15(a) & 1.85(b)

    Quantities are in ml.

Saline solution

10. Dissolve 8.5 g of sodium chloride in 1000 ml water: see pg. 57.

C. ANTIBACTERIAL SUPPLEMENTS

All these compounds must be obtained from a commercial source as a dry powder and made up fresh for experiments: see pg. 43.

1. Benomyl

    0.1 g should be suspended in 100 ml of sterile distilled water and added at 0.5% to the agar before pouring the plate to give a final concentration of 5 ppm.

2. Chloramphenicol

    Prepare the agar in the usual way, but just before sterilising add 0.5 g per litre of chloramphenicol to the boiling liquid (the antibiotic should be suspended in 10 ml of 95% ethanol/ethyl alcohol). Mix well and sterilise at 15 lbs. pressure per sq in (118^oC) for ten minutes only.

3. Penicillin

    Commercial penicillin is available in vials containing 60 mg of powder (= 1,000,000 units). One vial should be added to 200 ml of sterile distilled water and stored in a refrigerator for later use. Add 1.5 ml of this solution to 15 ml of cool medium just before pouring.

4. Streptomycin

    Commercial streptomycin, like penicillin, is also available in vials containing 1 g (= 745 units/mg). One vial should be added to 750 ml of sterile distilled water and stored in a refrigerator for later use. Add 1.5 ml of this solution to 15 ml of cool medium before pouring the plate, or use the solution directly after mixing 1.5 ml in 15 ml of sterile distilled water.

D. Micro-stains & Slide Mountants

Micro-stains

The recipes for micro-stains recommended for the accentuation of fungal structures in microscopic mounts are given below; some of the compounds are not obtainable directly from a chemist or drug store, and special arrangements should be made with the pharmacist, or the stain may be purchased directly from a biological supplier. Observations of tissues, individual hyphae, etc. are best carried out first of all simply by using a 10% aqueous solution of ammonium hydroxide. This reagent acts as a clearing agent and has a suitable refractive index to contrast with the fungal material. The solution is prepared by mixing 10 ml of ammonia (0.88 specific gravity) with 90 ml water.

Stain for nuclear protein (acetocarmine): see pg. 56.

Boil for ½ hour excess carmine stain (powder) in 45% acetic acid, prepared by carefully mixing 45 ml of glacial acetic acid and 55 ml water. After the mixture has boiled for approximately fifteen minutes, filter the resultant liquid and dilute it to half strength with 45% alcohol, prepared in the same way as the acid but substituting ethanol for the acetic acid. Add a drop or two of ferric chloride, prepared as on pg. 12 for agar media, to every 50 ml of reagent.

General stains

1. Melzer's reagent : see pg. 30, 56.

Iodine	1.5 g
Potassium iodide	5 g

Chloral hydrate	100 g
Water	100 ml

Add ingredients to the water, warm and stir thoroughly, but do not boil.

The stain is also used as a macrochemical reagent: see 'How to Identify Mushrooms III' pg. 25. In descriptions of agarics it is abbreviated to MR.

2. Cotton blue in lactophenol, or Amann's solution : see pg. 29, 76, 77, 78.

Add 20 ml of a 0.05% solution of cotton blue, prepared by adding 0.5 g cotton blue stain to 1000 ml water, to 45 ml of lactophenol. The lactophenol is prepared from the following ingredients:

Phenol (crystals)	20 g
Lactic acid	20 g (= 16 ml)
Glycerol	40 g (= 31 ml)

Lactophenol is a good clearing agent and therefore can be used alone as a mountant if the fungus material is pigmented.

3. Trypan Blue : see pg. 29, 76, 77.

Trypan blue can be substituted in the above formula for cotton blue as follows:

Trypan blue powder	1 g
Lactophenol	67 ml (prepared as above)
Water	20 ml

The strength of the stain often varies in both trypan and cotton blues, depending on the manufacturer's batch number. Comparative studies should be conducted by examining similar structures in the different mountants.

4. Congo Red : see pg. 56.

Congo Red	0.5 g
Ammonia (specific gravity 0.88)	33 ml
Water	66 ml

Thoroughly mix by shaking

5. Erythrosin : see pg. 38.

Erythrosin	1 g

Ammonium hydroxide 100 ml (freshly prepared solution)

The ammonium hydroxide is prepared by adding 10 ml of ammonia (specific gravity 0.88) to 90 ml water.

6. Phloxine : see pg. 76, 77, 78.

   0.5 g phloxine in 100 ml water
   Place the specimen in 70% ethyl alcohol on a slide in order to dampen the cells, and then run a solution of 5% potassium hydroxide under the cover-slip. Finally run an aqueous solution of 0.5% phloxine through the mount as outlined on pg. 29.

7. Cresyl blue : see pg. 60.

   Specimens are mounted in a saturated aqueous solution of cresyl blue which has preferably been filtered before use. The stain is excellent for distinguishing gelatinised tissues.

8. Giemsa : see pg. 55.

Giemsa	3.8 g
Glycerol	250 ml
Ethanol (absolute alcohol)	250 ml

   Thoroughly mix the stain in the glycerol and ethanol, and before use add 15 ml of phosphate buffer (see pg. 13) to every 1 ml of stain, so that a final pH of 6.9-7.0 is obtained.

9. Basic fuchsin : see pg. 52, 54, 56.

Fuchsin	1 g
Water (or ethanol)	100 ml

   See pg. 20 for Schiff's reagent which also uses basic fuchsin.

10. Carbol fuchsin : see pg. 56.

Basic fuchsin	1 g
Phenol crystals	1 g
50% ethanol	200 ml

    This reagent is also known as Ziehl-Neelsen's reagent.

Slide-mountants

1. Gum chloral

Gum arabic	40 g
Glycerol	20 ml
Chlorate hydrate	50 g
Water	100 ml

The mountant is prepared by dissolving the gum arabic in the water and adding the glycerol and chloral hydrate whilst heating the mixture in a water bath. Filter and allow to stand for about a week, but if stains are to be added this should be done before filtering. Specimens should be mounted directly in the gum chloral and allowed to dry. Temporary mounts should always be washed free of alkali before the gum chloral is introduced. The edge of mounts prepared in this way can be ringed with cellulose enamel or equivalent, or even nail varnish.

2. Polyvinyl alcohol

Polyvinyl-alcohol crystals	1.66 g
Lactic acid	10 ml
Glycerol	10 ml
Water	10 ml

Dissolve the crystals in the water and lactic acid whilst stirring vigorously; follow the polyvinyl crystals with the addition of the glycerol. Filter the product, and then leave for twenty-four hours before using. This mountant can be used directly on material stained with cotton blue in lactophenol or other acid dyes. Alternatively, cotton blue stain can be incorporated into the mixture during preparation at the rate of 0.1 g per 100 ml of mountant. Always warm the slide over an electric bulb or a spirit lamp to harden sufficiently for later examination under the oil-immersion objective of a light microscope. If mounts are left for twenty-four to thirty-six hours at $40^{\circ}C$ they will harden completely. Polyvinyl alcohol has an advantage over other mountants in its low refractive index.

3. Commercial Mountants

Euparal and Clearmount are commercially available mountants suitable as substitutes for Canada balsam: see pg. 29.

E. SECTIONING RECIPES (SEE PG. 52)

1. Fixative

   (a) Formol sublimate

Formalin	10 ml
Mercuric-chloride solution	90 ml

The formalin is prepared by mixing 40 ml of formaldehyde with 60 ml water and the mercuric chloride solution prepared by saturating a quantity of water with dry powder. (Mercuric chloride is very poisonous!)

(b) Alcoholic-acetic formal sublimate

Ethanol	45 ml
Glacial acetic acid	5 ml
Formal sublimate--solution (a) above	50 ml

2. Lugol's iodine

Iodine	1 g
Potassium iodide	2 g
Water	100 ml

Prepare by first dissolving the potassium iodide in the water and then following with the iodine: see pg. 53 and 54.

3. Alcian blue (see p. 53 for use)

    (i) Weigert's haematoxylin
       (a) Solution A

20% alcoholic haematoxylin	5 ml
Ethanol	95 ml

The haematoxylin solution is prepared by grinding the raw product into a mixture of 20 ml ethanol and 80 ml water, shaking up, and filtering. It is best to let the solution stand overnight before use.

       (b) Solution B

Hydrochloric acid	1 ml (concentrated)
Distilled water	95 ml
Ferric-chloride solution	4 ml prepared by dissolving 30 g of ferric chloride in 100 ml water.

    (ii) Acid alcohol

Hydrochloric acid	1 ml (concentrated)
Ethanol (70%)	99 ml

    (iii) Alcian blue

Alcian blue solution	50 ml

Dilute acetic acid   50 ml

The alcian blue stain is prepared by dissolving 1 g of the stain in 100 ml water; the acetic-acid solution is prepared by adding 1 ml acid to 100 ml water. Once the two solutions are mixed, the suspension should be filtered and 10-20 mg of thymol added to the filtrate.

(iv) Phosphomolybdic acid

Add 1 g of acid to 100 ml water

(v) Eosin

Add 1 g eosin dye to 100 ml water and thoroughly mix.

4. Periodic acid-Schiff reaction

Solutions required:

(i) Normal hydrochloric acid prepared by adding 98.3 ml of concentrated acid to 1000 ml water.

(ii) Periodic-acid solution prepared by adding 0.5 g of acid to 100 ml water.

(iii) Sodium-metabisulphite solution prepared by adding 10 g bisulphite to 100 ml water.

(iv) Sulphurous-acid rinse

Solution i above	5 ml
Solution iii above	6 ml
Distilled water	100 ml

(v) Carazzi's haematoxylin

Potassium alum	75 g
Haematoxylin	1.5 g
Glycerol	200 ml
Potassium iodate	0.3 g
Distilled water	1,200 ml

Dissolve the alum in the distilled water without applying heat. Grind the haematoxylin in the glycerol, and mix the two solutions so prepared together. Dissolve the potassium iodate in a small amount of distilled water and add to the mixture whilst shaking continuously.

(vi) Schiff's reagent

　　Basic fuchsin　　　　　　　　　　　1 g
　　Normal hydrochloric acid　　　　　20 ml (see above)
　　Sodium metabisulphite　　　　　　　1 g

The basic fuchsin must be added to the boiling water for dispersal, but before the acid is added the solution should be cooled to 50°C. Further cooling to 25°C is required prior to the addition of the bisulphite. The solution is not ready for use immediately; it should be stored in the dark at about 4°C until taking on a straw colour.

F. Preservatives (see pg. 24)

1. Formal-acetic alcohol: abbreviated FAA

　　Formalin　　　　　　　　　　　　　　7 ml
　　Glacial acetic acid　　　　　　　　7 ml
　　Ethanol (absolute alcohol)　　　　50 ml
　　Water　　　　　　　　　　　　　　　50 ml

2. 'Kew cocktail'

　　Methanol (comm. grad.)　　　　　　55 ml
　　Formalin (comm. grad.)　　　　　　 5 ml
　　Glycerol　　　　　　　　　　　　　 5 ml
　　Water　　　　　　　　　　　　　　　35 ml

Material can be left indefinitely in either of these solutions without injury until anatomical and cytological studies are made.

III. TECHNIQUES

A. Field Observations

1. COLLECTING TECHNIQUES

As with other fungi, it is always necessary to bear in mind when collecting agaric material for culture that the specimens should be suitable not only for laboratory work but also for later preservation as herbarium voucher material. The latter will then act as source material for further reference. Unlike microfungi where several stages of development are present on the same substrate and can be collected at one time, it is necessary with the agarics to collect a deliberate selection of fruit-bodies covering various stages from young buttons to the mature fruit-body. The material must be returned to the laboratory or home in as perfect and as fresh condition as possible. To achieve this it is best to carry a selection of tins and tubes and a supply of waxed paper; a basket is an excellent means of carrying this assortment of material. Always use plastic tubes to remove the danger of accidental breakage of glass ones by a fall.

The smaller specimens should be placed in the tubes or small tins, and the intermediate ones placed in tins or wrapped up in twists of waxed paper. The largest ones may be covered in the basket simply with grass or fern-fronds for added protection, or better still, placed separately in large paper bags. Plastic bags are not suitable; the fruit-bodies of many agarics perspire, and at the end of the day one is left with a soggy mess.

Specimens should be dug up complete, with the aid of a strong knife or fern-trowel, and placed in tins, or the more delicate specimens in tubes or waxed paper twists; for woody specimens, such as pleurotoid agarics and polypores, brown-paper bags are admirable. Remove as much soil and detritus from the base of the specimen as later cleaning may be difficult.

Wood-rotters should be cut off the wood carefully along with a small portion of the substrate; this will ensure that the basal structures are retained and allows the substrate to be re-determined at a later date if necessary. Like terrestrial specimens, they should be handled as little as possible. The identity of the substrate must always be noted, and if the fungus was growing on or under a tree the tree species, should be recorded. Notes on characters which may change on the journey back to the laboratory, or any which might be considered unusual at the time of collecting, such as small, fresh colours, and stickiness, should be recorded immediately.

Ecological notes are always useful, as reference to them may allow a previously stubborn fungus to be grown in culture or its spores germinated. Any ecological information can be incorporated into the design of the culture medium by using a fertile mind. Coprophilous fungi can be collected by taking some of the substrate, air-drying carefully, and then incubating in a damp chamber when time permits; the new crops of fruit-bodies will probably develop normally. Short notes on basidiomes, accompanied by a sketch (coloured if at all possible), are really necessary in order to make full use of the material; if in doubt as to the identity of a specimen expert opinion should always be sought. For cultural purposes, a fresh fruit-body in active growth is always preferable.

Examination of material

Once home, set out and examine the specimen carefully. A good spore print for determinative study can be obtained after one or two hours. Select a mature pileus from each collection, cut off the stipe, and set the pileus gills-edge down on a piece of white paper (if the specimen is small, on a microscope slide in a tin). Usually the smaller pilei give a satisfactory print soonest. A drop of water placed on top of small pilei is essential to prevent them from drying out. Even with large specimens it is desirable to place a glass slide between the gills and the paper; to prevent such specimens drying out, enclose in a tin or box, or cover in grease-proof or waxed paper (Plate 1; Fig. C & D). These same spore-prints can be later preserved. This is not the experimental material; the preserved material will act as a voucher collection. Whilst the sport-print is developing, field notes on the observable features can be made and the experimental work commenced; do not at this stage become involved in microscopy. The definitive identification using dried material and field notes can be made or confirmed at a later date by using the other volumes in this set. Indeed, if any identification cannot be achieved with your data, an expert will be able to help by examination of preserved material and the field notes. The chart found in 'Mushrooms I' will act as a useful guide so that important characters are not overlooked.

2. PRESERVATION OF VOUCHER MATERIAL

  a. Temperate countries

Agarics must be dried as quickly as possible in a warm room, or when an old-fashioned kitchen range is available, they may be left in the oven overnight for three or four successive nights after the fire has died down. Placing close to a central heating boiler is an alternative method. Too rapid drying makes the fruit-body brittle, but if not dried fast enough they will either be destroyed by maggots or become reduced to a soggy mass which finally dries solid. Laying the specimen on a radiator top is also very successful. With all techniques, allow for good air circulation, or one will finish with a 'casseroled' specimen which dries glassy. One should aim to have a source of heat with a rapid air exchange to draw off the moist air surrounding the specimen.

In the laboratory a commercial film-drying cabinet, although expensive, is excellent. It does not run at too high a temperature and it has good air movement. If a hot-air oven is used, then the door and vent should be left open or forced air circulation encouraged. Small bench heaters, or even an electric bulb with trays above, on which the specimens are placed are also good.

Smaller specimens can be dried by immersing them in a tube of silica gel; if an indicator is used then it can be seen when the gel is exhausted (pink) and must be replaced. The gel can be recharged by placing in a hot-air oven until blue again. It is useful to carry boxes of this material in the field, as important specimens can be treated immediately on collecting. They should be kept in air-tight plastic tubes, of which there are now many commercial brands available. Glass tubes should be avoided.

Larger specimens should be sliced into representative portions to facilitate even drying, although some curling will undoubtedly take place. However, slicing reduced the possibility of inner tissue becoming soggy and then collapsing, or more importantly becoming excavated by maggots. In the last case one will finish with only a shell of fungal tissue.

In the field it is more difficult to preserve material, especially if one is away from a source of electricity. When electricity is available, then portable driers can be used which utilising little power and can operate from a light bulb or small heater; a simmer switch can be attached just to keep the heat constantly supplied. A spare car battery will give a good source of electricity overnight, and on an expedition it can be recharged next day by replacing in the car. The same portable driers described above can be used with a lamp operated by some kind of liquid or gaseous fuel. Care must always be taken when carrying fuels such as this, and also when the heaters are being operated, as often space in a tent on an expedition is at a premium. However, such heaters are necessary when working in areas where it would be impossible to reach habitation by the evening.

A very useful household fruit and nut drier made by Sigg Dorrex of Switzerland consists of wire trays above a small porcelain heating element; the heat is circulated by a small fan operated by the convection currents produced by the element. Although designed originally for the kitchen, this drier is very light and easily packed for use in the field provided a source of electricity is available.

Having sent many collections to my home-base the double box system has proved the best. In temperate or cold climates, dried material should be replaced in twists of waxed paper after drying and arranged irregularly in a double-walled cardboard container made by constructing one cardboard box inside another. If one is lucky this can be achieved by simply placing a second slightly smaller box neatly inside the first, but more usually an inner box must be constructed by boards bent to that shape, making sure the sides and base are strong. When the box is filled, the specimens will lie irregularly and should then be compacted with light pressure from above, using the sheet of the cardboard to be used for the top. More specimens can then be placed in the box until again it is full. Repeat until the surface is level and there are no gaps between specimens, top board, and the top of the box. Never use strong force or artificial weights to close. Liberally sprinkle amongst the specimens flakes of p-dichlorobenzene or naphthalene to prevent attack by ants, beetles or mites. Wrap and dispatch. In more tropical countries the same procedure should be followed but the specimens should be placed in sealed polythene bags (or even two bags inside each other) with a little p-dichlorobenzene or naphthalene and silica gel. Do not rely wholly on the seal on so-called self-seal bags. If nothing else is available specimens can be laid between sheets of absorbent material soaked in a suitable preservative (see pg. 20) and sealed in polythene bags.

Material which may not be completely dried before packing should be placed not in polythene bags but in porous wrappings. Partially dried specimens are undesirable and should only be used when the specimens are really important, as moisture may be absorbed by the already processed material and so damage it. Always make doubly sure fully dried material is well protected from damp specimens, paper, packing, etc.

Spore-prints can be sent separately by air-mail so that cultures can be prepared as soon after making the collection as possible; the support material will follow. A stock of pre-sterilised strips of plastic the size of microscope slides is invaluable if wrapped in pairs or in packets of up to one dozen; they may be white, clear, or black. These can be used for making sterile spore-prints. By cutting one piece of plastic strip in half and placing partially under the cap of an agaric, the gills can be held vertically above the sterile surface. Hard glass or pyrex tubes should be used for pickled specimens; when cultures are prepared in the field this should be done in similar tubes. Each tube should be packed individually to prevent contamination or mixing of contents if breakage does occur. Tubes of pickled material should be filled with loosely packed cotton-wool or similar material. Only pack a few tubes at a time, in press-on lid boxes, for dispatch, and remember many countries have custom and excise restrictions on pickles containing alcohol. When polythene bottles are used, make sure they have good screw-on tops and strong shoulders to reduce distortion in transit. If possible the tops can be dipped into cool molten wax which is then allowed to harden.

All bottles and packets must be clearly labelled and always cross-referenced with field collection and note-books. Write in hard pencil or indian ink on good-quality paper but not on ply-card. Self-adhesive labels have been found to be of little use, and ball-point pens and water soluble inks should not be used. An abbreviated label system can be used if preferred, provided it is consistent and suitably documented in field note-books.

Specimens grown in culture can be preserved in the same way as indicated above. Delicate material can be dried between filter paper at room temperature. Always keep part of the spore-printed specimen in case the cultured and fruit-body material becomes mixed or lost.

It is essential in the preservation of all fungi including agarics to guard against attack by insects such as beetles or by mites, both of which can be very destructive to the collections.

If woody or corky agarics such as the pleurotoid forms are suspected of harbouring beetles, they should be enclosed for a few days in air-tight tins, such as a biscuit box, with a little carbon bisulphide. They can then be stored in boxes with a small quantity of naphthalene or p-dichlorobenzene. The use of the last two substances should be kept to a minimum, as in high concentrations they can have side effects.

Fleshy agarics nearly always harbour maggots when fresh. There are usually the larvae of mycetophilid flies which are feeding on the fresh fungi. They will die once dried. The greater danger to agarics is colonization by beetles after drying. A keen watch must be kept and action taken immediately when insect activity is noted. If the collection is large, fumigation with a commercially available chemical is necessary; small collections can be placed in tins and fumigated with carbon bisulphide or napthalene.

It is imperative to keep the areas in which the specimens are retained, dry. This discourages insect attack and prevents or avoids the specimens becoming colonised by micro-fungi and ultimately mouldy, and then unusable. The greatest difficulties will be encountered in tropical and sub-tropical areas where the atmosphere is damp for long periods. It is often necessary then to resort to pickling the collections. If adult flies or beetles are found in and/or on specimens, retain them for interested entomologists, as often the insects found are specific to particular fungi and their distribution is usually very poorly known. In some cases the adult stages of some of our mycophagous insects have never been described.

b. <u>Sub-tropical - tropical areas</u>

Difficulties in preserving agarics arise in warm climates mainly because of the humidity; unfortunately in many of the countries which experience this kind of climate facilities (including electricity) are often unreliable, even in cities. Remember also that generally the range of equipment so commonly accepted as part of the laboratory in the western world is usually not available.

All the equipment described in pgs. 22-25 above can be used in the laboratory or when on expeditions in the tropics. Although preserving fluid is bulky, its use is often less bothersome than drying specimens and then having frequently to carry out lengthy re-drying. Pickled material can be washed and slowly dried on return to one's home-base, or better still the material can be stored in the preservative (see pg. 20). If at all possible a single fruiting body, or part of it, should be kept in the dry condition. Collections may, however, be ruined if the dried material is not properly packed for dispatch back to the laboratory or home: see pg. 23-24.

Those preservatives which should be used are described on pg. 20. They can be kept in tightly fitting plastic containers or strongly sealed high-quality, coarse-guage polythene bags.

Material in the tropics should be processed as soon as possible and, once dry, cooled immediately, then transferred to polythene bags along with a teaspoon of naphthalene or p-dichlorobenzene and a dessertspoon of fresh silica gel. The bag can then be sealed by burning the top of the bag whilst it is held between two pieces of glass as if to make a sandwich (as shown in the diagram Plate 1, Fog. E). Do not staple any labels on to the plastic bag! Machines which seal plastic bags are now commercially available.

Always remember that some countries impose restrictions on the importation of plant material and also on the importation of quantities of alcohol. ALWAYS make the necessary enquiries before you visit a foreign country on an expedition or collecting trip.

B. LABORATORY

1. ROUTINE LABORATORY TECHNIQUES

When the basidiospore, asexual spore, or phphal fragment is allowed to grow on an artificially produced food source under controlled conditions, the growth subsequently produced is called a <u>culture</u>. The food source is called the <u>medium</u>. Such cultures can be produced in any suitable container, but a flat dish-like structure with a loosely fitting lid (Plate 1, Fig. B) is traditionally used; it is called a <u>Petri dish</u>. Cultures produced in this way are the soft-ware of not only the experimental agaricologist but the mycologist in general and the bacteriologist. The Petri dish in which the fungus is growing (indeed any container in which a fungus or bacterium is cultivated) needs to be sterile in order to eliminate the development of contaminating organisms. These organisms may be detrimental to the growth of the agaric being studied, or even if not detrimental they are a nuisance factor, spreading from one culture to another by producing either clouds of spores or slimy masses of cells. Such contaminants compete with the material under study for food. Glass Petri dishes can be sterilised by the following methods:

1. Heating to $150°C$ in a hot-air oven (domestic gas or electric cookers are quite suitable), preferably packed in tins or individually wrapped in aluminium foil, and leaving overnight.
2. Rinsing out with alcohol and drying off by flaming.
3. Fumigating by leaving in them overnight small pads of cotton wool soaked in propylene oxide; the pads should be removed and the dishes aired before use.
   If Petri dishes are not available, the squat jars with screw caps in which household produce is packed can be used. Their lids should, however, be loosened to allow air movement.

After the dishes are sterilised, melt the medium by placing a vial in a saucepan of boiling water. When molton allow it to cool to about $45°C$. Flame the caps and mouths of the vials by passing them through the flame of a spirit lamp or bunsen burner before pouring the medium into dishes. DO NOT lift the lids off the dishes too high but only sufficiently to permit the mouths of the vials to be introduced; try not to touch the edge of the dish with the vial. Allow the dish with its contents to cool in a clean area.

<u>Slide culture</u>   Plate 1, Fig. 1

The vegetative stage and sporiferous structures are often extremely fragile and can be destroyed simply in the attempt to transfer them from a culture plate to a slide. Techniques are available to facilitate their undisturbed growth and examination. Such cultures are called <u>slide-cultures</u>.

Procedure: Pour a plate of selected agar medium as described previously (pg. 26) and allow the agar to set. Then cut two slits in the agar with a previously flamed scalpel or other suitable instrument as shown in the diagram (Plate 1, Fig. A & B). The flaps of the agar formed by the cuts should then be lifted and a microscope cover-slip inserted under each; return the flaps to their original position. The dish should then be inverted and marked over the middle of the cover-slip squares with a grease (chinagraph) pencil. These markings show through the agar when the dish is reversed. Cut out pieces of agar corresponding to the squares. The medium should then be inoculated in the centre of the plate by taking a small piece of flesh or mycelium or small group of spores on the end of a sterile needle (as described below) and placing on the agar surface under as sterile conditions as possible.

The fungus will grow over the exposed area and can be examined with the low power of a compound microscope or a binocular microscope. When sufficient growth has been made the surrounding agar should be cut out and the cover-slip removed and mounted for microscopic examination. In order to have both hands free for manipulating the scalpel or the mounting needle, etc. it has been found that binocular magnifying goggles are excellent; These goggles can be purchased quite cheaply from a discount store or mail-order firm.

Another method of preparing a slide culture is to place a small drop of selected agar medium 5-7.5 mm in size onto a slide, previously sterilised by dipping into alcohol and passing through a flame (Plate 1, Fig. B). Inoculation is then carried out and a similarly flamed cover-slip placed on the top. The slide should then be mounted in a damp chamber prepared as outlined below, so that growth can take place unhindered, although monitored by periodic inspection. The fungus grows out from the medium in the narrow space between the cover-slip and the slide. After sufficient growth has been made the cover can be removed and the medium picked off and discarded. Some of the specimen will remain adhering to the glass slide and some to the cover-slip. Both slide and cover-slip can be used to prepare a microscopic mount for examination.

Alternatively, the slide with both agar and organism colonising the medium can be gently warmed over a spirit-lamp until the agar melts and the cover-slip by its own weight flattens the culture.

Damp chamber Plate 1, Fig. B

A damp chamber can be made quite easily from any closed container previously sterilised inside with alcohol or as directed on pg. 27. A Petri dish is of course ideal, or for larger numbers of slide-cultures a plastic sandwich box obtainable from the supermarket is admirable.

The bottom of the dish should be covered in wet, sterilised blotting paper or cotton wool and the slides supported by small pieces of flamed glass rod, glass rings, or pieces of glass slide laid on the paper. The chamber is then ready for use.

Slide preparation

Normally a drop of microscope stain, the most important ones of which are described on pgs. 14-16, is placed on the slide and a small waft of the fungus from a culture picked up with a sterile needle and placed in the drop. The slide should then be examined under a dissecting microscope, binocular microscope, or binocular viewer and the material teased out and arranged using needles. Separating the hyphae is best carried out in either a 10% aqueous solution of ammonia (see pg. 14) or 5% solution of potassium hydroxide in water, although clean water is quite adequate. The slide and cover-slip cultures described above should be treated in a similar way, by adding a drop of stain and a cover-slip to the former, then lowering the latter slowly onto the drop of stain on a clean glass-slide.

Sometimes dry spores carry air-bubbles with them and so it is useful to put a drop of dilute teepol or similar commercial washing-up liquid on the fungus first. The excess liquid should be drained off before mounting as outlined above. A small piece of scotch tape pressed onto the culture, mounted directly in the stain on the slide, and viewed provides a very quick method of obtaining a superficial idea of the structure in culture of the agaric being studied.

Again with those species producing dry spores a pre-treatment with teepol or ethyl acetate is adviseable to disperse air-bubbles. Air can be dispelled from a mount by placing a drop of alcohol at the edge of the cover-slip and drawing the liquid through with a piece of blotting paper or filter paper.

Slide preservation

If one wishes to retain the mounts for later observation, one can do this by using commercially available nail-varnish. Indeed any old unwanted nail-varnish belonging to one's wife, daughter, secretary, or assistant will do; it should be thinned by the addition of a little amyl acetate and thoroughly stirred. Remove any excess stain from the slide with absorbent paper or blotting paper and lay a piece of the same over the slide. Apply gentle pressure to the edge of the cover-slip with a finger until all excess moisture has been removed, taking care not to move the cover-slip. A narrow band of nail-varnish painted around the edge of the cover-slip and on the slide, and allowed to dry and harden, will stabilise the mount. This band anchors the cover-slip so that the slides can be gently rinsed with water to wash off any excess reagents. The moisture should be removed with the application of a piece of blotting paper or similar absorptive material; the slide should then be set aside to dry. To make a final seal, apply a broad band of cellulose-enamel cement over the nail-varnish.

A very useful mixture for preserving temporary mounts is gum chloral, as this is water soluble and dyes can easily and to advantage be incorporated into it. The preparation of gum chloral is described on pg. 17. Cotton blue and trypan blue stains prepared as outlined in the recipes on pgs. 15 can be added to the gum chloral mixture before mounting the specimen. Warming, but not boiling, the mount over an electric bulb or a spirit lamp will harden the gum chloral sufficiently for examination of the mount under the oil-immersion lens of a microscope. Full staining will take several hours, and mounts left for 24-36 hrs at 40°C will harden completely. The stain should be added to the gum chloral and thoroughly mixed to give a uniform mixture. Melzer's reagent can also be used with gum chloral.

Euparal and Clearmount are commercially available mountants which can be used as substitutes for Canada balsam. Polyvinyl alcohol is just as good and can be made quite easily in the laboratory from the recipe indicated on pg.17.

Comparative studies

For comparative studies the following procedure is recommended to give reproducible results.

Grow the isolate under study on 2% malt extract agar in a Petri dish for one week; the medium should be made up as directed on pg. 8 . From an actively growing culture, cut cubes two or three millimetres square, and transfer them to the edge of each of five Petri dishes containing 20 ml of malt extract agar. Incubate the cultures at room temperatures in the dark; expose the cultures to the light only for examination. Describe all the macro- and microscopic characters once a week, every week, for six weeks. When cutting out small pieces of the culture for examination, make sure that the scalpel has been sterilised by flaming, each time and every time. Examination of all five cultures allows characters to be recorded with ease without there being any need for doubt.

In contrast, when one wishes to assess a culture's ability to produce extracellular oxidase systems, the following procedure has been traditionally used. More sophisticated techniques are described on pgs. 86-89.

Procedure: Petri dishes containing gallic (or tannic)-acid agar, prepared as directed on pg. 4, should be inoculated in the centre of the dish. The onoculum should be four or five millimetres square and should be taken from an actively growing culture four to six weeks old. A positive reaction will result when a colourless zone is formed about the onoculum; activity can vary from weak to strong, or may be totally absent.

References for more sophisticated techniques and refinements of this method will be found in Appendix IV.

2. TECHNIQUES FOR SECURING PURE CULTURES

Although comparatively simple to carry out, culturing requires a considerable amount of patience and care. Careless work can do untold damage, whereas careful attention to hygiene can allow good work to be done even under the most appalling circumstances. The main aim should be to reduce hazards of accumulation of dust and draught during culturing by installing close-fitting windows and doors and keeping the furniture in the room to a minimum.

Before use, the bench on which the culturing is to be done should be washed down with disinfectant, and the glass-plates on which the inoculations are to be carried out should be rubbed down with absolute alcohol and flamed; surfaces of laminated plastic should only be swabbed. Fifteen minutes or so before inoculating, spray the air of the room with a mild disinfectant, eg. thymol in 30% ethanol, which will precipitate any floating spores. The floor should be regularly disinfected. A dampened towel can be effectively laid on the bench top in dusty climates, large enough to cover the working area and contain the rack supporting the sterilising dquipment. A commercial electronic precipitator, although an additional expense, has been shown to be very useful in clearing the air in a room of unwanted spores.

When using a spirit lamp for flaming, however controlled, realise this is still a dangerous operation.

In tropical to subtropical areas, drape any openings in the room or tent with damp muslin or cheese cloth, and cover the table with a wet towel.

Small inoculating hoods may be constructed quite easily to provide an enclosed area for transferring cultures. Such structures should be sterilised before and after use with disinfectants, absolute alcohol, or ultra-violet radiation.

    a. <u>Isolation from basidiome (fruit-body) tissue</u> Plate 3, Fig. A-C.

       (i) <u>Isolation from hymenium and pileus (cap)</u>

Just as the most successful choice of media for growth of fungi depends on the species, even on the individual, used, so the area from which the tissue should be selected for culture may differ in it's growth. The purpose of the basidiome, ie. fruiting body of the agaric is to produce basidiospores, and although some parts of that basidiome may die during maturation or become dormant, that area which is most active for the longest period is often the actual spore-producing tissue. The basidia and related structures are produced on the hymenophoral trama. In the agaricoid fungus this is frequently the best place to seek the tissue plug which will be utilised as an inoculum. The basidiome should be cut, or preferably broken open, trying not to touch any inside surface, under as sterile conditions as possible. A piece of tissue 3-5 mm square should then be taken with a flamed scalpel from the central area of the gill, tube-wall, or fold, or where it joins the main flesh of the fungus. With the woodier pleurotoid specimens, it may be necessary to cut away the external skin with a scalpel first, then with a second flamed scalpel cut out a small piece of tissue. In soft, watery specimens, bacteria and other fungi are invariably present. Agarics are often slow-growing in culture, and these bacteria and invading fungi may grow and swamp the inoculum.

Select material which is not overmature, to reduce risk of taking material infested with other fungi, bacteria, or yeasts. It is preferable to culture the same day, particularly if the basidiomes are old; younger ones may be kept in the refrigerator overnight. It is really unnecessary to sterilise the pileus and stipe surfaces with iodine, Chlorox, or similar disinfectants, if sufficient care is taken not to drag the needle with the inoculum across the outside. If required the specimen can be pinned down to a board or better still stuck with 'blue-tac' to a previously sterilised glass plate. Half-immerse the fragments of tissue in the selected agar, and grow in the dark.

Several mycorrhizal species die very easily after isolation and need constant attention; they should be cultured immediately after collecting.

Sometimes the hyphae do not grow out from the block if placed on solid agar, particularly if the tissue has been taken from a mycorrhizal agaric or bolete. However, when the tissue is floated on the edge of a small piece of sterilised paper on a nutrient solution, hyphal filaments can be observed.

(ii) Isolation from stipe (stem) and velar fragments

Some agarics have little or no pileus tissue, and extraction of the smallest piece of flesh would mean also taking along tissue which had been in contact with the environment; in such cases stipe tissue can be a very suitable substitute material. The stipe of the basidiome, when present, acts both in water transportation and support for the usually more delicate apparatus producing the basidiospores. Thus there are areas of the stipe which may be quite dead but others which are in an active phase of growth.

With a flamed scalpel the fresh basidiome should be cut or broken to expose a clean inner surface. A small block of tissue should then be taken from the stipe-apex from just within the cortex, or if hollow away from the central air-space, or from the stipe-base where active connections exist with the substrate. Care must be taken with the last inoculum, as other fungi and bacteria may enter the basidiome from the soil, growing up along with the host hyphae.

With more delicate fungi, the stipes can be either plated out directly onto agar or washed in vials of water, treating them as if they were rhizomorphs (see pg. 33). The hyphae of the stipe will proliferate on plating out. Contamination appears to decrease or even disappear upwards from the basal section. However, a continual surveillance is necessary to make sure contamination does not take place.

Many agarics possess a veil which protects the developing spore-producing layer. Fragments of this veil from an actively growing fungus can be treated as delicate stems and plated out directly onto agar, or if more leathery treated as if segments of a rhizomorph. The inner surface of the veil in an agaric in which the pileus rapidly expands, is very suitable, as are the filamentous cells of certain coprini; see pg. 110.

b. <u>Isolation from vegetative phase</u>

(i) <u>Isolation from soil</u>

Techniques using dilution plates and hyphal isolation surprisingly have succeeded in the culture of several agarics, particularly those with sclerotial bodies or other resting stages; see pg. 73. Sclerotia can be obtained either by actually digging amongst the bases of fructifications or by sieving for smaller ones, or by utilising helminthological (nematological: eel-worm) techniques. Active mycelium can be obtained from these resting bodies by slicing or by transferring the softer medulla to agar, eg. with *Lentinus*. Surface sterilization, or burying the sclerotium in sand flushed with nutrient solution or embedding it in nutrient agar, may also be successful. Techniques for dilution plates are described on pg. 39.

Rhizomorphs or mycelial fragments can be picked directly from the soil. If dry shake free of soil debris, or if moist wash in sterile water. It has been found advantageous to treat the rhizoids of agarics as if they were mycorrhizal short roots, i.e. wash up to a dozen times in small quantities of sterile water in vials before cutting into segments and finally plating out.

Vegetative mycelium can be found attached to the base of the basidiome, and in an actively fruiting population this is frequently a very reliable place from which cultures can be obtained. Washing up to a dozen times in small quantities of sterile water in vials has been found adequate, and surface sterilization with a 0.1% aqueous solution of hydrochloric acid for 10 seconds gives very favourable and consistent results. Dipping in a 10% aqueous sodium-hypochlorite solution of even domestic bleach for a few seconds gives very adequate results.

(ii) <u>Isolation from wood-samples</u>

Many agaricoid fungi can be directly isolated from wood-specimens, provided the samples are taken from the edge of the decay where the hyphae are active and in profusion. The wood-sample should be either washed with 10% aqueous sodium hypochlorite for 10-20 minutes or with 1:1 (w/v) mercuric-chloride solution, flamed lightly, or washed with any commercial mild sterilising fluid used for cleaning baby's bottles or other glass-ware. Treatment with a 1% aqueous solution of silver nitrate followed by a rinse in a 1% aqueous solution of common salt (NaCl) also has proved successful.

Rinse thoroughly with sterile water after treatment with all reagents above before proceeding. A clean surface of the decay should only then be exposed with a flamed scalpel, and a small piece of wood cut out and placed on a nutrient agar. The mycelium may take some time to develop, so the sample should be left for at least four weeks. More sophisticated techniques of cutting wood and taking samples have been developed for other basidiomycetes, but the above is sufficient for agaricoid fungi.

(iii) <u>Isolation from mycorrhiza</u>

For ectotrophic mycorrhizal fungi, a group which includes a large range of agarics and boleti, the short roots of the tree are washed several times in distilled water and are cut into portions and plated out as described for rhizomorphs above.

Isolation from roots can be accomplished simply by dissecting the roots into two parts, an outer cortex and an inner stele, in sterile water. These are either plated out directly or incorporated into nutrient agar after further fragmentation; dispersal in the latter case is carried out by shaking and rotating the plate before the agar solidifies. Examination should be made regularly and hyphae isolated as soon as growth is noted. Ectotrophic mycorrhizal fungi are often deficient in thiamine, so a glucose/ammonium tartrate or (Modess') glucose/ammonium chloride medium should be used if growth is poor (see pg. 5).

It may be found ultimately that some agarics prove to be endomycorrhizal in parallel to other basidiomycetous fungi. Endotrophs can be isolated by washing plant roots in a solution containing 10% hypochlorite solution for 10-20 minutes, or in a 1:1000 mercuric chloride solution (HgCl) for 2 minutes. Wash the roots further in sterile water to remove disinfectant, and finally cut them into two or three segments before plating out aseptically. Mycelium will grow out and can be cut off with suitable manipulation and cultured. Coils of the endophyte can actually be taken out of roots by lightly squashing under pressure with the flat edge of a scalpel.

c. <u>Isolation from spores</u>

(i) <u>Basidiospores</u>

All techniques described above produce dicaryotic colonies, but often it is preferable to possess the monocaryotic culture (see pg. 92). This can be achieved by utilising the meiosporic stage (≡basidiospores) of the agaric. The conventional way of using such a spore is to prepare a suspension of the spores in sterile water. Suspensions so formed can then be spread with a loop onto the agar surface. Alternatively, the entire vial of diluted spores poured onto the agar surface and the Petri dish turned to spread the liquid evenly. In this case the dishes are then left for a few seconds, after which the excess liquid is poured off, being careful not to dribble the liquid down the lip of the dish. Whilst viewing under a dissecting microscope, pick off with a slender needle any germinating spores, preferably when the germ-tubes are still quite short.

Although many aspects of spore germination have been reviewed, the clear fact is that we still need to know a lot more about the physiology of these fungi. Only with great difficulty have some spores been induced to germinate, and spores of some even common agarics, eg. <u>Pluteus cervinus</u>, have resisted all attempts. It is impossible to indicate all the possible conditions which have been and must be tried to induce germination, but they include cold and heat treatments, proximity to hyphae of the same or a different taxon, proximity to other microorganisms, presence of specific metabolites, presence of plant extracts in the medium (see pg. 5), and incorporation into the medium of activated charcoal. The last is often useful in order to induce mycorrhizal agarics to germinate, as it apparently removes one or more self inhibitors.

As the percentage of germination is still often poor, careful vigilance is necessary. It is advisable to pick off the germinating spores as soon as they grow. Spore-prints not required immediately can be stored dry if kept in a cool place, eg. refrigerator at $4.0°C$.

The following techniques can be recommended:

a. Place the spores of the fungus under study on a piece of sterilised cellophane or sausage-skin. This membrane should then be placed spore-side up directly onto a culture of the same or a closely related agaric. When the spores germinate pick them off.
b. In artificial culture spores may germinate, and indeed agarics fruit, if they are in close proximity to contaminating colonies of saprophytic bacteria or yeast. It is on occasion useful to introduce these organisms artificially into the pure culture to encourage spore-germination. Apparently, essential growth substances are released into agar by the contaminants, and these compounds stimulate dormant spores to germinate.

c. Plant extracts. A range of media which can be used to encourage spore germination has been given previously, but really any vegetable or fruit can be substituted in the recipes provided. Thus prune, and chick-pea (Cicer) agars have been found particularly good for stimulating germination.
d. For coprophilous and other dark-spored agarics, pre-treatment in an air oven at $60^{\circ}C$ for 1 hour of $37^{\circ}C$ overnight before plating out often induces germination.
e. Pancreatin. Many coprophilous fungi pass through the gut of an animal during their life-cycle. Treatment of spores by washing in an aqueous solution of pancreatin has been found to parallel in some way the changes in environment normally experienced by coprophilous fungi.
f. Furfuraldehyde. A 0.01% aqueous solution of furfural added to the molten agar prior to pouring into the Petri dish has been found to give encouraging results. Washing the spores in a 0.001% solution of the same reagent prior to streaking out on agar has also been successful.

As we know so little about the germination of spores, it is difficult to advise as to which medium should be used for a particular group of agarics, but ecological observations in the field can narrow the possible choices to one or two media which should be tried first.

Good results have been obtained by removing a plug of agar from a previously poured and cooled agar plate and filling the hole so formed with a solution of a selected medium, eg. the culture medium made up without the addition of agar-agar. The nutrients will diffuse out from the well into the agar, producing a mosaic of food concentrations which parallels the spectrum or gradation of food materials found in natural substrates. The wells can be made with a flamed cork-borer; spread the basidiospores of the agaric under study over the agar surface before removing the plugs (see pg. 22). Cork-borers often have a lip at their base; this should be filed down to remove any ridges which could transfer unignited alcohol to the culture.

A double-gradient agar plate has also proved of great use. Although originally designed to analyse the effects of antibiotics on bacteria, it can be applied to studies on spore-germination. By pouring half the usual quantity of agar medium used into a Petri dish propped up with a small block, a slope can be formed. After solidification, the Petri dish is leveled and an equal quantity of tap-water agar added to the dish. The plate is left for twenty-four hours to allow diffusion of nutrients to commence. Spores can then be spread onto the surface of what is simply a graduated medium.

Fresh turgid spores which have been collected from mature basidiomes and immediately plated out on malt extract agars (2; pg. 3) often germinate after a 48-hour period when maintained at $25^{\circ}C$ in the dark. Spores of many species of agarics, however, have not germinated satisfactorily in culture, even within 28 days. From a single species, many collections of only one or two spores have been observed to develop vesicles and germ-tubes. The subsequent growth of these germ-tubes has been poor, and the vesicles have soon collapsed. Germination of these species did not improve even if spores from fresh material were agitated in sterile distilled water for ½, 1, 2, 3 or even 4 hours. Sterile dilute aqueous solutions of organic acid and/or mineral nutrients also have no apparent stimulatory effect. These results suggest that the lack of germination is not due to the activity of a simple water-soluble and leachable inhibitor.

Spores from herbarium material of many agarics, however, germinate if they are incubated in a saturated atmosphere overnight, transferred to a vial of sterile distilled water, dispersed by mechanical agitation in a micro-shaker or shaken by hand vigorously, and plated out.

Incubation is best carried out by placing the dry spores in a welled slide mounted on glass disks in a damp chamber, as described earlier (pg. 28). The slide should be surrounded by dampened blotting paper or suspended above a 5mm depth of sterile water. This chamber should be left overnight in a warm darkened incubator; $27^{\circ}C$ would appear to be most favourable temperature. By this simple technique, basidiospores taken from material maintained for many years in the dry conditions of an unfumigated herbarium have been induced to germinate. Washing spores off the gills of dried specimens is sometimes successful. Basidiospores from herbarium material treated with insecticides have not germinated under this method.

A very satisfactory technique for overcoming the induced dormancy in spores once they have escaped the basidiome is to eliminate completely any drying-out processes. A section of the freshly collected agaric pileus is cut with a group of gills still attached to the tissue in such a way that it can be suspended above the surface of a selected agar medium. This can be done in a Petri dish or similar container or in a McCartney vial or test-tube containing an agar slope. In the first case the pileus tissue is stuck to the dish lid with a smear of vaseline. The dish lid can be rotated over a 12-hour period so that a deposit of spores is made over a good proportion of the agar surface. Once a deposit is produced the lid should be replaced by a clean sterile lid. When using the vial or tube, cut a small wedge of the agar slope off the thin end, then rotate it on the inside of the container and push it downwards to rest immediately above the portion of the agar slope remaining. This procedure is shown in the diagrams (Plate 2 figs. D-J).

The piece of pileus with gills attached can then be introduced and left for 1-2 hours, or more if need be until a spore-print is obtained. After this period the pileus and small piece of agar should be removed and discarded. Any contaminant, even mites, will remain on the small 'island' of medium and thus a clean spore-print is obtained, the basidiospores being discharged from the specimen onto the agar surface below. The spores can be left to germinate, or mixed with distilled water and streaked, or sucked up in a syringe and sprayed out onto agar. When the agaric is small, the whole pileus may have to be used in the technique, but if the agaric is thick-fleshed, a piece of the pileus flesh immediately attached to the gills need only be used. The benefit of these techniques is that the spores are not allowed to dry out, as they fall fully turgid from the basidia onto the agar.

Spores of most coprophiles and many lignicolous fungi germinate immediately by this technique. Some do not, possibly because of the presence of self-inhibitors produced by a high concentration of spores. In this latter case the spore-print should be cut out, agar and all, and mixed in a micro-blender; the debris should then be plated out, distributing the spores over a wide area. If the spores are easily wettable, it is necessary simply to wash the spores off the agar with sterile water.

(ii) Asexual spores

Agarics produce asexual stages far more frequently than many mycologists appreciate, and such structures can be used to advantage in culturing. Asexually produced spores can be either dicaryotic or monocaryotic. One of the asexual processes found in the agarics is chlamydospore production, and this can take place either on or in the basidiome or on the vegetative hyphae. Frequently the chlamydospores are large enough to be picked off from a drop of sterile water. Other asexual spores are known, some formerly being termed oidia; there is little doubt even more will be recognised in the future. These asexual stages appear already in the schemes of classification of the Fungi

Imperfecti and will be described later pg. 66.

Adopting similar techniques to those described above for basidiospores, one can treat the easily removeable veil cells of certain agarics (eg. *Cystoderma, Phaeolepiota,* and *Coprinus*) as propagules; we are uncertain whether these structures ever act as propagules in nature. In these fungi, the powdery surface from the pileus epithelium should be rubbed across the agar and allowed to develop further.

Many basidiomycetes inhabit soil, and it is only now becoming obvious that techniques employed by microbiologists and helminthologists (=nematologists) can be of use to those who study agarics.

d. <u>Indirect methods</u>

Techniques for isolation of fungi from soil have been described in many books on the culture of micromycetes; see Booth, refs. in Appendix. Those relevant are described below:

(i) <u>Soil dilution plates</u>

(a) Place a small quantity of soil in a tube containing 10 ml. of sterile water and mix vigorously with a clean sterile pipette attached to a bulg. Draw off 1 ml. of this solution and add to a second tube containing 9 ml. of water. Repeat the process at least twice, then mix the resultant liquid with molten agar cooled to about $40°C$ and pour into a Petri dish.

(b) Alternatively, melt five tubes of selected agar and allow to cool to about $45°C$. Inoculate one tube with a small amount of soil and mix thoroughly by rolling the tube between the hands. Then pour the plate. Refill the empty tube with another tube of agar, roll and pour to make a second plate. This procedure should be repeated.

(ii) <u>Soil plates</u>

Break up a small sample of soil (0.5-0.75g) in a sterile Petri dish, spread it across the bottom, and knock off any excess particles by tipping the plate. Melt a selected medium and cool to $40°C$ before pouring onto the remaining soil.

(iii) <u>Flotation method</u>

Mix a small sample of soil with a small amount of mineral oil and shake vigorously in water. Plate-out the emulsion which collects on the surface.

In the three techniques outlined above, fungi other than the one required also grow on the culture plate. These contaminants should be removed, or the isolate required should be transferred to a clean plate. Transference of an isolate is carried out with a flamed dissecting or inoculation needle (Plate 1, fig. B). The agar should be cut so that part or all of the colony is included in a small block. After resterilizing the needle, impale the small block and transfer it to a clean agar surface. Contaminants which may accompany a colony may be similarly cut out very carefully and, without shaking, transferred to and immersed in a dish of absolute alcohol. Alternatively, a drop of absolute alcohol can be dropped from a micro-pipette or syringe directly onto the contaminant.

(iv) Hyphal isolation

Many hyphae adhere tightly to soil particles. After a soil sample is shaken in water, the heavier particles separate out from the suspension; these can be collected with a sieve of 50 μm pore diameter if considered necessary. Examine this residue microscopically for the presence of fungal hyphae which should then be removed with a previously sterilised needle or similar tool. Transfer to a Petri dish and cover with cooled medium.

3. TECHNIQUES FOR GROWING AGARICS IN PURE CULTURE

   a. Inoculating

      (i) Vegetative tissue

Nichrome wire bent into a hook at the tip and mounted in a handle makes a good inoculating needle. Immediately before use it should be sterilised by passing through a flame, then allowed to cool before use. The metal cap of the vial bottle in which the culture is growing should be flamed before it is opened.

A small tuft of mycelium should be removed with the needle and planted on an agar surface in a Petri dish which has already been poured and cooled. To reduce the risk of contamination by airborne spores, the dish should be opened only sufficiently to permit the insertion of the needle. All observations can be seen through the clear top of the dish. Naturally the culture should be handled in a clean area, and, if possible, in a laboratory in a sealed inoculation chamber or 'clean' area cut off from the rest of the working space or on a commercially available flow table.

Sometimes it is only necessary to touch the medium at one spot with the loaded inoculation needle or to make a streak. Because many agarics are restricted in their growth, it is usually better to inoculate at several points. Always monitor the growth of the fungus.

(ii) Basidiome or Fruiting body tissue

A small piece of tissue should be cut out of a basidiome with a scalpel and the process above repeated, but on this occasion always inoculate at various points on the agar surface, preferably in a concentric zone about 30 mm. from the margin of the dish.

(iii) Basidiospores and Mitospores

Spores can be directly picked up in groups from a spore-print with a nichrome wire moistened in sterile water. Mitospores (asexual) can be picked up similarly from the surface of the culture. The spores should be directly streaked or spotted onto the medium, or transferred to a vial of sterile water and then shaken vigorously to disperse the spores before plating onto the agar surface. When using a suspension, allow the plate to stand for a short while before pouring off the excess. Take great care not to dribble any residual liquid down the outside, or to allow any to run back onto the agar surface. Alternatively this same suspension can be streaked onto other poured plates.

The germination of the spores must be monitored, and any producing germ-tubes should be picked off with a nichrome wire. If this is not possible, the spores should be located with a microscope and positioned in the centre of the microscope field. The objective should then be racked up and replaced by a 'biscuit-cutter' objective previously sterilised by passing through a flame. On racking the cutter down on to the agar, a plug will be cut out of the culture; this should then be transferred to a fresh agar surface. It is sometimes advantageous, when high contamination is likely, to mix the spores in a solution containing 30 µg streptomycin or similar broad-spectrum antibiotic before plating out; see pages 13-14.

b. Cultivation and Maintenance

(i) Procedure after isolation

Once the hyphae have begun to grow out from the inoculum (tissue, wood-sample, root or other source), they should be cut off and placed on a clean plate. In the case of spores, it is best to pick out several which are germinating and allow them to intermingle on fresh agar; this will ensure that an anastomosing colony with sufficient genetic information to form a balanced secondary mycelium is produced. Because in the majority of fungi the apical cells contain more than one nucleus, only the tip of a single hypha is required (four or five cells) in order to give a balanced thallus. The colonies so produced can be grown on general or specialised media and the colony examined periodically for sectors and contaminants. "Clean' areas 5 mm^2 can then be cut out and maintained as a pure stock. Always check the morphology of the spores of the original basidiome with that of the germinating ones to ensure that the two belong to the same species.

Watling (1968) has reviewed the various media which have been used for isolation of specialised groups of basidomycetes, but it is usually quite sufficient to isolate the fungus on a potato-extract agar (eg.PDA, PCA) or an oat-meal or malt-extract agar (1.5-3%). In all cases the agars should be made up in the laboratory from raw material in preference to "made-up" media, and only when these substrates have failed should more specialised media be considered (eg. Peptone/dextrose malt agar, MDA, Pg. 3; YA, Pg. 4). Incubation conditions should fall into line with the ecology of the fungus, in parallel to the suggestions above.

(ii) Restrictions and purification

Clamp-connections characterise the basidiomycetes, or so the textbooks inform the student, but there are many agarics which lack them. Only experience can teach the investigator how to recognise the mycelium belonging to a basidiomycete when clamp-connections are absent. In contrast, many mould contaminants can be recognised by virtue of their characteristic fructifications. Several fungi, eg. *Chaetomium globosum* and *Gymnoascus* spp., appear to reach maturity very slowly, and it may be several subcultures after the primary isolation before they become apparent and fruit. A few species of mould are frequently isolated from the tissue of agarics, eg. *Calcarisporium*, *Fusarium* and *Trichoderma viride*, and obviously are present therein as parasites. What is more surprising is that recent observations indicate that even the hyphae of different species of agarics may parasitise one another. It is therefore doubly necessary to ensure that the right fungus has been isolated, particularly as perennial basidiomes, even though they are still growing, may be permeated by their own mycelium or by the mycelium of an alien fungus. This phenomenon is apparently not confined to woody fungi; records of intimacy between hyphae of unrelated basidiomycetes in the field are now common place, eg. *Suillus bovinus* and *Gomphidius roseus*, *Rhizopogon parasitica* and *Brauniellula nancyae*, *Armillaria mellea*, and *Entoloma abortivum*. In parallel, some *Coprinus* spp. have been isolated in the laboratory from the stipe-tissue of other unrelated species of the same genus. Although from a taxonomic view-point this is extremely perturbing, the fact still remains and has now been reproduced several times. It is, however, quite understandable, when one can observe that when some coprophilous fungi are grown together in culture they do not separate out one from the other; the advancing front of the colony is then composed of the hyphae of more than one species.

In general, agarics produce a rather fluffy to silky colony frequently aggregating in areas to form strands or knots of mycelium. Such fluffy colonies are particularly characteristic of wood-rotting fungi. Several coprophiles, however, produce a submerged, greasy-moist colony with very little aerial growth.

Cultures prepared from basidiospores of some species of agaric frequently exhibit two distinct types of colony, i.e. a fast-growing colony and a small slower-growing colony. These appear to be inherent characters of the fungus.

For control of bacterial contamination, several suggestions have been made over the years ranging from the addition of inorganic salts to the medium to the use of antibiotics, or more simply to a change in culture technique. Wood-decaying fungi are frequently able to decompose (and utilise) phenolic compounds, some breakdown by-products of which are toxic to bacteria and other fungi. It was with this ecological fact in mind that a medium containing 0.006% of σ phenyl phenol or 0.004% α napthol has been designed on which to grow basidiomycetes.

It is best, if possible, to do without the assistance of antibiotics in case any slow change in the mycelium is induced, for unlike the ascomycetous fungi we are ignorant of many aspects of the physiology of the basidiomycetes. However, sometimes antibiotics must be used, and a measure of success has been obtained by incorporating Rose Bengal (0.135g/litre) or potassium tellurite (0.1g/litre) into the culture. Streptomycin and similar broad-spectrum antibiotics have also been used, and recently benomyl has been added to the range of suitable compounds (see page 13). A more advanced manual should be consulted if further information is required.

Inoculation of the basidiomycete onto sterile wet wheat straw will in most cases allow rapid separation of bacteria and fungi because of preferential colonization of the substrate by the organisms; the fungus can be reisolated from the frowing front. A further way to separate bacteria and fungi is to allow the mycelium to grow around a glass tube onto the medium, or to cut the colony out and place the mycelium face down on a fresh agar surface. The mycelium in the first case will grow over the tube and if held horizontally 2-3 mm from an agar slope will spread onto the surface of the agar block by bridging the gap. In the second case the mycelium will grow through the agar block more rapidly than the bacteria are able to colonise the block. The fungus can be then isolated free of contaminants.

There are some isolates which although commencing healthy and active become 'lazy' and show signs of hyphal degeneration. These appear in gross characters similar to cultures of cultivated mushrooms attacked by virus. Such cultures are best discarded, but techniques of heat treatment, which may be applicable to them in the future, have recently been developed to free mushrooms of virus.

4. TECHNIQUES FOR MAINTENANCE OF CULTURES

Subculturing is best carried out onto wooden blocks, into sterile dung or other suitable material after approximately six months; the choice of substrate depends on the ecology of the original isolate. Cultures are best stored at $4°C$. Transfer from an old culture to a new one is always ultimately necessary, as when cultures are maintained on agar for long periods there appears to be a fall-off in mycelial activity and ability to fruit, perhaps because of lack of certain growth factors. The carbon/nitrogen ratio of the media on which agarics are maintained is often critical. Mycorrhizal fungi are generally difficult to maintain on semi-synthetic media, and the addition of vitamins, such as thiamine or one of its constituent moieties, is often required. Other techniques employed, eg. use of mineral oil to preserve cultures, parallel in every way those adopted for Ascomycetes and Fungi Imperfecti.

The most favourable temperature at which agarics grow and fructify in culture varies with the species. Thus *Coprinus cinereus* grows well at $27°C$, whereas *Armillaria mellea* and *Lepista nuda* fruit only after growth at low temperatures ($10-12°C$). *Coprinus cinereus*, for instance, often grows on dung heaps, which may heat up considerably from bacterial activity, whereas in Europe *Lepista nuda* and *Armillaria mellea* require cooler temperatures, appearing late in the year after the first frosts. Most agarics grow well at $22-24°C$, and those that commonly fruit in the laboratory do so between $20-24°C$; remember temperatures for vigorous vegetative growth may be quite different to the temperatures which favour fruiting.

Although the cultivated mushroom is often grown in darkened caverns, the majority of agarics grow best in good light. Incubators or cabinets in which the cultures are retained should allow as much light through as possible. A well illuminated, clean window-sill is admirable for those without laboratory facilities.

It is not recommended that cultures should be retained for long periods, owing to the increasing possibility of 'weed moulds' entering the cultures, infestation by mites, or degeneration of the fungus, with time. The maintenance of stock cultures of fungi can satisfactorily be carried out only in correct working conditions under the supervision of experts. Cultures should always, for instance, be stored away from dust and damp.

Synthetic media prepared from only salts and carbohydrates are not recommended for long-term preservation, as fungi grow neither typically nor vigorously when on them. Media based on fruits and plant-extracts contain all the substances, including suitable trace elements, which are required for growth, and are therefore recommended. Some organic carbon source is essential, but if it is supplied as sugar alone, only vegetative growth is stimulated and the cultures soon become sterile. The carbon source should preferably be a mixture of organic compounds.

Potato/carrot agar (PCA) is a very favourable medium on which to conserve cultures. If agars containing large quantities of free sugars are used for isolation (pg. 3), the fungus should always be transferred to a medium less rich in sugar for retention in a culture collection. Cultures should be re-inoculated every three months unless special equipment for freeze-drying (lyophilization) is available. McCartney vials are one of the best kinds of containers for maintaining cultures over long periods. Cultures on agar slants should be grown until they produce a healthy development of mycelium, then they should be covered to a depth of 10mm with sterile mineral oil (Pharmacopoeia quality). This technique is excellent in tropical countries, as it prevents the drying out of the cultures, discourages penetration by mites, and requires no expensive apparatus. Again McCartney vials, with their rubber liners removed, are ideal.

Except when in store, cultures in oil are rather messy to deal with, and therefore other procedures have been introduced. One such technique is to use cigarette papers and an adhesive. Procedure: flame a cotton-wool plug and push it down into the tube containing the culture, flame again, and while the rim of the tube is hot, push the tube's mouth into a mixture of 2 g of copper sulphate in 20 g of gelatin and 100 ml distilled water. Whilst still hot, the tube should be held tightly onto the centre of half a cigarette paper previously sterilised with propylene oxide. Hold firmly until set, when the surplus paper can be flamed off to make a tidy job.

Lyophilization is a sophisticated technique demanding fairly expensive equipment and skilled operators. The attraction of such a technique is that it uses only a small sample and protects the cultures from attack by mites. The culture is placed in a glass ampoule and dried in the frozen state whilst under a vacuum. The ampoule is finally plugged and sealed, also under vacuum.

Before the fungus can be recultured, the ampoule must be broken open.

Liquid nitrogen is certainly more hazardous than any of the methods described above, but it is excellent for keeping cultures of fungi. It is based on a technique of suspended animation developed for keeping animal-sperms and bacteria. This method has been very successfully applied to many agarics.

For effective storage in the college or home laboratory, cultures should be kept in vials in a cool place to reduce their rate of growth; a 4-8°C domestic refrigerator is admirable. For more details of the more sophisticated technique available see reading list provided.

All agarics are susceptible to attack by mites, small invertebrates related to the spiders and harvest-men; The mites will invade tubes, bottles and Petri dishes and eat the hyphae, infect the culture with bacteria and yeasts, and by wandering from one culture to another will carry contaminants. They thrive particularly well in high temperatures and high humidity, and so are more common in areas with tropical or warm climates; in temperate countries they thrive in summer.

Mites are attracted by the odour of the fungi and can enter the laboratory either on the bodies of flies, on plant-debris, on human beings, or on fresh basidiomes brought in for culturing. The methods of controlling mite attack are numerous, but few are entirely successful. Crude tractor vapourizing oil has been used to repel mites, and vaseline containing methyl benzoate smeared on the tubes is effective. Paradichlorobenzene crystals enclosed with cultures in a box probably kill any mites present but this is mildly toxic to both fungus and humans. General hygiene is the best way of preventing attacks by mite.

5. TECHNIQUES FOR INDUCING FRUCTIFICATION   PLATE 3

Some agarics are known to fruit directly on the Petri dish from basidiome tissue or from a spore-inoculum. Others fruit erratically, sometimes even producing fructifications in the stock tubes, but this is a very haphazard event. Successful fructification of agarics has usually been confined to dung-loving (coprophilous) or wood-inhabiting (lignicolous) species. In order to induce fruiting, media which contain natural extracts, eg. vegetable (pg. 5), yeast, malt, or dung, have always been found more successful than chemically defined media. No single method can be expected to induce fructification in all the agarics, for they behave very differently, one species to another, even in their natural habitats.

Flammulina velutipes, *Schizophyllum commune*, and species of *Agrocybe*, *Coprinus*, *Panaeolus*, *Psathyrella*, and *Psilocybe* produce normal basidiomes in culture; others, however remain sterile or produce abnormal or abortive fructifications (see pg. 74). Many fungi, including mycorrhizal species, produce fruit-body initials, but these cease growth for inexplicable reasons. A common feature of many fungi is that the original isolate fruits readily whereas subsequent subcultures do not. Abnormal fruit-bodies is some agarics, particularly pleurotoid species, are induced by dry conditions prevailing in the culture-vessel.

When culturing some *Coprinus* spp. from stipe-fragments, basidiomes often appear directly on the stipe-tissue, whilst other *Coprinus* spp. fruit directly from the pad of hyphae produced from a spore or group of spores. In yet other species of *Coprinus* fructification is observed only when bacteria are also present in the cultures.

In general, for the successful production of fructifications, relatively high humidity and a supply of rich, moist well-aerated medium is required, with exposure to light of moderate intensity. The magnitude of each Parameter must always be judged from the ecological notes one has made on the fungus in the field.

Vermiculite kept within porous, hollow, soft tiles and moistened with nutrient solutions has been employed for the fructification of *Schizophyllum*, but this is generally a very easy fungus to fruit. An effective, although simple, procedure in order to maintain a fairly high humidity in the culture is to cut a disc of agar from a colony in a Petri dish and fill it with either water or a nutrient solution, eg. dung or malt extract. The same technique has been suggested earlier for inducing spore-germination (pg. 35 & 36).

If fructification commences in the Petri dish and the lid is interfering with growth, it can be carefully removed and replaced by a pre-sterilised deeper lid (perspex domes commercially available as Gro-more seed-kits are ideal). The dome allows the basidiome to develop to its full maturity without the walls of the vessel fouling its development.

It is well known that some fungi require fairly high temperatures for growth (thermophiles), whilst others grow quite well vegetatively at room temperatures. Similarly some fungi require heat treatment to induce fructification, eg. *Coprinus delicatulus* (= *C. cinereus*), whilst others require a cold shock, eg. *Flammulina velutipes* and *Pholiota* spp., *Armillaria mellea* and *Lepista nuda*. It cannot be over-emphasized therefore that ecological studies must run parallel with any cultural studies.

Several fungi, particularly *Coprinus*, will fruit on plates or in flasks containing semi-synthetic media after the addition of some natural extract which has been decided upon by consideration of the ecology of the taxon. In nature, the growth substance coprobin, required by many coprophilous fungi for growth, is present in dung. However, as far as many species of *Conocybe*, *Coprinus* and *Bolbitius* are concerned, the dung can be replaced without ill-effect by an extract of soil. Members of *Coprinus* section *Setulosi*, like species of *Conocybe*, *Psathyrella* and *Panaeolus*, fruit well on Lange's dung agar (pg. 4). The best results appear to be obtained with media of rather loose consistency by preparing the agar with a gel concentration of not more than 1.5%. Media with the addition of dung extract and peptone, a mixture of potato/carrot (+ dung), corn meal (+ dung), or malt extract (+ dung) have all been utilized with success.

It may be advantageous under rather special circumstances to treat the culture with black light, ie. radiation peaking at 3650$\overset{\circ}{A}$ and with a range 3100-4100 $\overset{\circ}{A}$. The technique has been used successfully with many micro-fungi for stimulation of fruiting, but of course it is a sophisticated technique available only to those in laboratories. Its full use in the study of basidiomycete cultures has not been realised.

Often fructifications develop unevenly on the agar surface in the Petri dish, undoubtedly because those primordia which appear first monopolise the food material and draw nutrients from other, younger primordia. This physiological response can be reduced by cutting the agar into sectors; indeed cutting blocks out of the agar often itself induces fruiting. In liquid cultures, basidiomes are often initiated at the junction of the liquid meniscus and the glass vessel; indeed in many experiments primordia form in large numbers there and allow the full range of developmental stages to be found (see page 59). Liquid cultures are obtained by inoculating bottles containing media prepared in the same way as indicated in the recipes described in pages 3 - 7 but lacking agar-agar.

Unfortunately, whole groups of the higher fungi have failed to fruit in culture, but on considering physiological and ecological aspects of a single taxon or group of agarics there is little doubt many of their mysteries will gradually be revealed. Some of the initial work is already available in the literature as physiological exercises, for such procedures are necessary preliminaries to later studies which may be designed with a more genetic than systematic bias. Bearing this in mind culture-systems have been especially designed to induce fructification, and media have been prepared to simulate something resembling the soil system more than the familiar, inhospitable plaque of agar in a Petri dish. Thus moist shredded, loosely interwoven paper-pulp incorporated with nutrient agar offers to the fungus cavities containing air, pockets of liquid, and plugs of solid medium. The thin layers of nutrient agar covering the pulp are quickly colonized, apparently supplying the necessary amount of food required for successful fructification. This medium was designed for terrestrial fungi as an extrapolation of an artificial substrate originally suggested for lignicolous species; the latter substrate consisted of sawdust impregnated with an accelerator prepared primarily from bone and maize meal. Etter's medium consisting of a mixture of corn meal, corn starch and powdered wood has also been used with some success for terrestrial agarics; it too was originally designed for wood-rotting fungi (see page 10).

The procedure for the preparation and use of the paper-pulp medium described on page 10 is given below as an example and guide for those wishing to grow agarics on solid media. The paper-pulp medium has been chosen because it has given consistently reproduceable results.

Place pads of previously soaked, shredded paper-pulp in half-pint bottles to a 10 cm depth and partially squeeze out any excess water. Cover the pad with sufficient nutrient agar so as to form a 1 - 2 cn layer at the bottom of each bottle and a thin coat over the paper-pulp. The agar can vary depending on the ecology of the agaric under study, although malt-extract agar is worth using first to monitor growth. The alveolate nature of the pulp presents a large aerated surface for fungal colonization. With a flamed cork-borer, cut out discs of agar on which either spores are germinating or mycelium is growing, and inoculate the bottles. The bottles should be kept on a clean window-sill at $20^{o}C$.

After about 50 days primordia appear, and within 60 days mature basidiomes bearing normal basidia, basidiospores, cheilocystidia, and dermatocystidia are formed. The spores produced by these fructifications are viable and fresh inoculations have been made from them. Some cultures, however, take longer to produce primordia, or once they are formed they develop no further. Favourable growth and the production of basidiomes in these cultures can be induced by covering the paper-pulp with a layer of a sterile sand and soil mixture, similar to that called 'casing soil' in mushroom-culture (see page 98) under *Agaricus*). The paper-pulp can be a rag, ie. cellulose waste containing some cotton, or a hardwood or softwood derivative. Three more types of paper-pulp which have been utilised are a *Eucalyptus* hardwood sulphate pulp, a softwood sulphate pulp and a softwood sulphate unbleached pulp. The first of these three has been found to give the best results; it is characterised by its short fibres. The second pulp is characterised by long fibres and the last by its high content of lignin. Whereas many pulps stratify whilst breaking up the raw pulp in a macerator, the eucalypt pulp rapidly disintegrates during maceration to give a very uniform easily manipulatable material, which reacts well to squeezing and later expansion, both important when preparing the medium.

The pulp can be replaced by whole cereal seed or maize-kernels. If such a medium is preferred, then the maize or wheat should be soaked overnight in the liquid medium chosen as being most favourable to the agaric under study. To produce large numbers of basidiomes of coprophilous agarics, sterile dung of an appropriate type is preferrable.

To induce fructification in some species of agarics, a supply of casing soil, either natural or sterilised is required; this procedure is normal practice in cultivating mushrooms.

Soils containing high quantities of organic material, when sterilised often produce toxic compounds which can inhibit growth. Such toxic material, particularly volatile substances, are also known to accumulate in cultures and may either discourage fructification or even inhibit vegetative growth, or both. In contrast some volatile by-products of growth have been shown to be stimulatory. All these compounds may be removed by forced aeration. Activated charcoal added to the pulp during the preparation of the growth bottles described above removes one or more of these toxic vapours and self-inhibitors.

About 50% of the European species of *Coprinus* fruit quite easily after growth on agar media in an incubator at $24^{\circ}C$ for 7-10 days; the same cultures would produce similar results after longer periods if left at room temperature ($20^{\circ}C$) in the light. Some species fruit much better, however, if one to two rabbit pellets are added to the agar. Cultures formed after plating out macerated fragments of mycelium often fruit within one week, because in this technique the whole plate is seeded with hyphal fragments. Sclerotia from the wild will usually fruit if placed on sterile coarse sand which is slightly moistened with water or nutrient solution.

Strangely, in some species fruiting will occur significantly if about eight inocula of each mating type are introduced onto each plate, whilst a single inoculum of the appropriate dicaryon of the same species may never fruit (see genetic studies page 90). In species with blocker mutations (see page 93), basidiomes tend to occur first along the junction line between the dicaryon and monocaryon and then ultimately elsewhere on the plate. Some species fruit on the inoculum plug itself, whilst others form a ring at a short distance from the inoculum.

Ectotrophic mycorrhizal fungi are even more coy to fruit than the most difficult of the coprophilous agarics, and it is usually more by good luck than good management that those which have fructified in culture have done so; the experimental techniques available at the moment do not appear to give reproducible results in pure culture. Some researchers have achieved the fruiting of mycorrhizal fungi by exposing plants grown under sterile conditions to the spores or mycelium of the fungus-associate under observation, producing the short infected roots and then waiting patiently for primordial development.

As the mycorrhizal tree must be considered a dual organism, every attention must be given to the physiology and ecology of the tree as well as the agaric. An attempt must be made for the tree to be irrigated with nutrient solutions, prefereably according to a regime reflecting the seasonal pattern of growth. Thus by the construction of a suitably designed tank, nutrients can be given to and withdrawn at will from the root-system of the tree and residual nutrients flushed out.

To carry out such a study, tree-seeds should be washed for 3 mins in 1:1000 w/v mercuric chloride ($HgCl_2$):water suspension, or more preferably in the less toxic and commercially available Chlorox. They should then be placed in a sterile clean Petri dish to allow the chlorine to dissipate before placing on the surface of water agar where the seeds will germinate. The seedlings when they commence to grow should be carefully eased from their seed-coats and the 'plantlet' placed on a freshly prepared agar surface fitted with a plastic dome (see page 47). As many contaminants are embedded in the seed-coat, the treatment is not sufficient to kill all the fungi, and by hastening the emergence of the seedling, contamination can be avoided. The seedling when large enough should be transferred to a vermiculite/peat-moss or sand/peat-moss mixture in Erlenmeyer flasks or similar containers.

This technique has been successfully used for *Lactarius rufus*, *Laccaria laccata*, *Paxillus involutus*, and *Hebeloma* spp., and for the gastromycetes *Pisolithus tinctonus* and *Scleroderma citrinum*, members of a group of fungi very difficult to grow, let alone fruit. A similar technique is used in ignorance in the semi-artificial cultivation of truffles by peasants in Southern France.

Primordial initials, which normally fail to develop on synthetic media, have formed in these cultures and have continued to grow after special care. Shock waves can either induce or disrupt primordial development, and atmospheric content and flow, quantity and quality of light and its periodicity all can play important roles. Most of these parameters have been poorly investigated. Some of the boletes which have been taken to completion in culture are ecological border-line cases in families which, although characteristically mycorrhizal, include fungi which are distinctly humicolous or lignicolous, eg. *Boletus sulphureus* and *Boletus lignicola*. Although primordia have been seen in several species, only for a few boletes have they been taken to maturity, but fruiting even in these is very inconsistent. Only once has a true *Suillus* been successfully taken to completion in the laboratory. In the Boletaceae and Russulaceae there appears to be a graduation of interdependence on the tree relationship, ranging from less specialised to truly obligate mycorrhizals. It is at the former end of this scale, e.g. *Lactarius rufus* and *Boletus subtomentosus* and *B. amarellus*, that success has so far been achieved. Perhaps others will be cultured with ease in the coming 20-year period.

C. STAINING TECHNIQUES

For many years it has been assumed that the structures which constitute the basidiome of agarics are all derived from the same tissues and pass through the same developmental patterns and phases. This is certainly not so. The "mushroom-shaped" form so familiar to every field mycologist is derived through 10 well-differentiated pathways; equally the surfaces of the pileus, of the stipe, etc, are constructed not only in different ways but often from quite different building blocks in one group of agarics to another. Examining stained sections of developing basidiomes with histological reagents reveals these differences clearly. Before the tissues can be examined, sections must be prepared. Small blocks 3-4 mm^3 should be cut from the basidiome, fixed in chemical preparations to prevent deterioration, and stained with specially selected dyes. The reagents used in these techniques are described on page 17-19. Three blocks should be sufficient for each separate study. The area of the basidiome selected should be chosen so as to demonstrate:

    i LS of pileus to show pileocystidia, pileipellis, structure of cortex, trama and subhymenium, context (flesh) characters, and hymenial elements.
   ii TS of tubes or lamellae (gills) to show hymenophoral trama and hymenial elements.
  iii TS of stipe to show caulocystidia, structure of cortex, context, and veil.

1. PROCEDURE

    a. <u>Fixation</u>: Use 20-50 times the volume of fixative to fungus material, and leave for 4 days. The specimen can in fact be left in fixative for anything up to 4 weeks without deterioration; for longer storage transfer to 80% ethanol prepared by adding 80 ml of ethanol to 20 ml water.

    Vials of fixative can be kept in the laboratory or taken into the field. Suitable field notes describing the fresh basidiomes must be taken before the material is placed in the fixative. If possible, always keep in addition to the experimental material another basidiome or two to be dried and processed as herbarium specimens as outlined on pages 22-24.

    It is very useful to add a drop of a 1% aqueous solution of basic (or acid) buchsin to give the block of tissue some colour; in this way the tissue can be more easily located in later treatments.

The following schedule should then be carried out:

    (i) <u>Dehydration</u>

(a) Dehydrate in 80% ethanol overnight
(b) Place in ethanol (absolute alcohol) for 2-2.5 hours
(c) Repeat twice with fresh absolute ethanol, each treatment for 2-2.5 hours.

    (ii) <u>Clearing</u>

Clear in either chloroform or carbon tetrachloride overnight.

(iii) <u>Wax impregnation</u>

Immerse in parafin wax (54-56°C melting point) for 2-2.5 hours at 58-60°C. Repeat the procedure twice more with fresh wax.

(iv) <u>Cutting sections and mounting</u>

Sections should be cut about 5 μm thick on a microtome and mounted on albumenised slices. The sections can then be dried in an incubator at 56°C, never for less than 2 hours and preferably overnight. Once prepared, these sections can be kept indefinitely. The albumen mountant can be purchased already prepared from a biological supplier and should be thinly spread on the slide. Before the slide is ready, the albumen should be allowed to dry.

(v) <u>Rehydration and clearing</u>

Remove the wax by placing the slides supporting the sections into xylol for 2 minutes. They should then be placed for the same time in ethanol (absolute alcohol), followed by a further treatment of 2 minutes with a mixture of 70 ml ethanol and 30 ml water.

After this time the sections should be washed in water for 20-25 minutes. The mercury deposit which may be present because of fixation can be removed by treating with Lugol's iodine for 3 minutes followed by a rinse with water. Now the iodine must be decolourised with a 5% solution of sodium thiosulphate for 2 minutes.

b. <u>Staining</u>: Many staining schedules are available and can be consulted by turning to the relevant zoological and histological texts which cover stain-technology. Fungi do not stain well with the usual range of botanical stains, and the following schedules have been found the most useful for studying development.

(i) <u>Alcian blue staining schedule</u>:

Before commencing the staining schedule, mix equal parts of solutions A & B outlined in pages 18 (i) to prepare the fresh haematoxylin stain which will be used below.

Stain the sections with Weigert's haematoxylin for 10 minutes, after which they should be washed for at least half that time in water. Differentiation can be achieved using the acidified ethanol for 3-4 secs before washing again in running water for 10 minutes. Rinse after this washing with distilled water. Then place for 10 minutes in the alcian blue solution made up to the recipe (iii) on page 18. Rinse again and treat with solution (iv) for 10 minutes. Then stain with solution (v) for 10 minutes after this treatment. The sections should then be transferred to a bath with running water and washed for 5 minutes. Then dehydrate, first in 30% ethanol for 5 mins, then 50% ethanol for the same time, then 75% ethanol, and finally absolute. Mount in canada balsam or in resin, or any mountant of one's choice.

(ii) Periodic acid-Schiff's staining schedule

Rehydrate and remove mercury deposits with Lugol's iodine; rinse sections with water and then treat for 5 minutes with 0.5% periodic acid solution (outlined on page 19). Rinse again in distilled water, and place in Schiff's reagent, also as prepared on page 19, for 30 minutes. The sections should then be rinsed in 3 changes of sulphurous acid, each change being of 2 minutes duration. Wash in running water for 10 minutes, and stain the nuclei by immersion in solution (v) for 1 minute. Wash the sections immediately in running water for 10 minutes, clear, and mount in appropriate mountant.

2. ALTERNATIVE SCHEDULES

Excellent results are obtained by the techniques described above, although they are usually considered laborious and old-fashioned. Sections do not have to be embedded in wax but can be cut on a freezing microtome. With a freezing microtome, sections should be cut at thicknesses of between 5 and 10 μm. Float off these sections, or flick the sections up with a camel hair brush and place in distilled water containing a drop of 1% aqueous basic fuschin stain to colour the sections. The same dye is used in the schedule above when blocks are taken from the original basidiome: see page 52.

Carbon dioxide can be used to cool the stage of the freezing microtome. If so the specimens are best mounted in gum-arabic solution, prepared as described on page 17 but lacking the chloral hydrate. A commercially available attachment (Pelcool) consisting of a thermocouple-fitted stage replaces the carbon dioxide method, removing the necessity to store gas cylinders. If this system is used, then the specimens should be mounted not in gum arabic but in 10% glycerol prepared by mixing 10 ml of glycerol with 90 ml water.

Failing the availability of sophisticated techniques, a sharp razor blade (either a commercial disposable kind or an old-fashioned cut-throat razor) can be used with pith, carrot, or potato as outlined in 'How to Identify Mushrooms III'. Although all the tissues will not be seen clearly with this simple technique one will be able to demonstrate that in different species not all rings on the stipes of basidiomes are formed from the same tissues, that not all hymenophoral tramas are similar, and that the cortex of the pileus is not directly comparable one species to another.

3. SCHEDULE FOR STAINING NUCLEI AND RELATED STRUCTURES

Fix the material in the same way as outlined above, be it hyphal growth or basidiome tissue. Rinse the sections in 95% ethanol, prepared by adding 5 ml of water to 95 ml of ethanol, and then transfer for 1-2 minutes to 70% ethanol prepared by mixing ethanol and water in proportions 7:1. The sections should then be rinsed in water and immersed in N hydrochloric acid for 5 minutes: see page 19. Remove the material and rinse in clean acid for 7 minutes at $60^{o}C$. Rinse in water and repeat the washing 5 times more, before transferring to a suitable phosphate buffer to give pH 6.9-7.0 (see page 13) this treatment should be repeated with fresh buffer at least 5 times. Place in Giemsa stain, prepared as outlined on page 16, after this treatment. It is necessary to remember to add a buffer to the stain before use, as advised on page 13. The slides should be treated for 2-3 minutes in the stain, washed first in buffer and then in water, and then air dried. Mount as directed on page 54.

The staining time may have to be modified slightly for different species of agarics; a few trials should be run and the results assessed. Nuclei are usually very difficult to stain, and great patience is required to obtain good results. Slowing down or speeding up the reaction with the hydrochloric acid (hydrolysis) may have drastic results. Trial and good judgement are required!

Nuclei can also be stained by using the wide-spread acetocarmine stain, a routine tool of the cytologist; it is prepared as outlined on page 14. Procedure: Take a piece of basidiome tissue or weft of vegetative hyphae and mount in a drop of the reagent on a glass slide. Heat gently over a spirit-lamp until the merest trace of steam is seen, but do not boil. Agitate the tissue with an old scalpel blade, mounting needle, or similar rusty object (the rusty instrument supplies sufficient iron ions to intensify the reaction). When the droplet has evaporated, repeat the process twice more with fresh acetocarmine. For the final agitation, transfer to a clean slide, heat as above and then place the slide on a cold surface after warming. A cover slip should then be placed on the material, and if necessary a little more liquid run under the slip. Nuclei will stain with a dark purple-brown colouration in much the same way as they do with haematoxylin in the staining schedules above.

Exactly the same technique can be used to view carminophilic granules or, as they are better termed, siderophilic granules; these granules characterise certain members of the Tricholomataceae and Entolomataceae. The material used for this exercise must be a maturing gill or tube, but the technique can be applied to fresh or even herbarium material. The siderophilic granules if present appear as numerous dark spots scattered within the basidia.

Siderophilic granules should not be confused with fuchsinophilic granules which are structures found on the hyphae constituting the pileipellis of many species of *Russula*. Here the granules take on a dark appearance when mounted for 15 minutes or so in a strong aqueous solution of carbol fuchsin. The excess stain should be washed off with a 10% solution of hydrochloric acid prepared by adding 10 ml of the acid to 90 ml water. The tissue before examination should be transferred to a fresh drop of the same acid solution for one minute. Finally, mount the tissue in water and examine.

Basic fuchsin can also be used to stain nuclei and siderophilic granules. It is known as Feulgen's reaction, and the procedure is outlined in most elementary botanical text-books.

Staining for pigmentation

Holland's picroformol reagent has proved to be the best formulation to accent pigmentation of hyphal walls and so allows one to assess the distribution of pigment within a tissue. The technique is described in 'How to Identify Mushrooms III': After treating the basidiome tissue with the fixative for 5 minutes, the material should be agitated in concentrated chloral hydrate and brought to the boil. Before microscopic examination it is best to transfer the specimen to a fresh drop of concentrated chloral hydrate.

Dolipores, the 'trade-mark' of the Hymenomycetes, are minute structures incorporated into the septa of the hyphae (plate 10 fig v). They can be accentuated by immersing in a drop of Congo Red stain prepared as directed on page 15. The same structures in *Boletus calopus* can be viewed by mounting a piece of basidiome tissue in Melzer's reagent.

D. Genetic Techniques

Genetics is the study of the mechanism and patterns of breeding of organisms and is very demanding in its experimental approach. Very accurate techniques have been developed to study the genetics of agarics, and some of the more simple ones are described below. In another section of the book (pgs. 90-97) the field is expanded still further.

1. SINGLE-SPORE ISOLATION

Transfer a mass of spores, about enough to cover a pin-head, to a vial of sterile, distilled water and shake up vigorously. Tip the contents onto the cooled surface of poured agar in a Petri dish, and rotate gently. Make sure the lip of the tube, which should be flamed prior to pouring, neither touches the Petri dish side, nor any dribbles are allowed to run down the vial onto the agar surface. Allow the Petri dish to stand for a short while and then pour off the excess liquid, again being careful not to allow any dribbles to run back onto the agar surface.

Under the low power of a microscope, single spores can be observed and picked off free-hand or with a mechanical device (eg. a biscuit-cutter attachment to the microscope which replaces the objective as outlined on pg 41). A microprobe can also be used; such an instrument can be purchased already complete or made from small pieces of excess bricabrac as outlined by Kemp. A reference to the method of making such an instrument is given in the Appendix pg. 125. The spores once isolated should be transferred on the needle to new agar surfaces in either tubes or dishes. The cultures so produced are called monocaryotic.

Single-spore cultures can also be prepared from basidiospores which have been allowed to fall onto an agar surface set just below a sporulating surface. Tetrads (the four spores which surmount a single basidium) will fall close together and can be picked off on a piece of agar as a unit of four; they can then be separated further as outlined above. When a piece of tissue with gills is placed on a glass slide and observed under a dissecting microscope, maturing spores will be seen; tetrads ready for dispersal will attach themselves to the point of a fine needle placed close by. In this way the tetrad can be transferred to an agar surface and the individuals separated. Allowing the gill to dry on a glass slide in the open laboratory will give a poor spore-cast, but some of the spores produced will be in tetrads, and these can also be picked off with a needle and placed directly on agar.

2. MACERATION (PLATE 11, FIG. A-E.)

Macerating a colony, inclusive of the agar, followed by incubation of the plated fragments, has been useful in 'cleaning' cultures of agarics from contaminants, but it is equally a method of producing monocaryotic isolates. Procedure: Add 4 ml of sterile saline (pg. 13) or sterile water to a thick-walled glass (test-tube) tissue grinder which has been previously sterilised in alcohol and flamed. A 3-mm block of the culture under study should be placed on the tip of the plunger as illustrated in the diagrams (Plate 11 Figs. D-E), and then pushed down six to eight times without rotation. The culture disintegrates, and the fragments so formed should then be poured out, taking the precautions outlined under single-spore isolation above. The dishes are then placed in an incubator at $22^{\circ}C - 25^{\circ}C$, and when any fragments begin to elongate they are picked off, free-hand or with a manipulator, and transferred to fresh agar surfaces. Some of the fragments will have lost their protoplasm in the treatment and will die, but others will grow to form active colonies.

3. DE-DICARYOTIZATION

Maceration is a mechanical way of separating the nuclear types contained in the mycelium or basidiome tissue of a basidiomycete and is called de-dicaryotization. The same phenomenon often occurs spontaneously in the laboratory when cultures are maintained for long periods under artificial conditions; it may also occur in nature in response to environmental phenomena. By macerating a colony, chlamydospores, if present, will also be released (see pg. 68). If these are separated and allowed to germinate, hyphal tips can be cut off when produced and transferred to clean medium. Colonies so formed may be either monocaryotic of dicaryotic, generally the latter. Chemicals have also been used to encourage de-dicaryotization by incorporating various substrates into the agar. 1.5% sodium cholate may be used, with the culture then held at $25°C$ for one month. The hyphae which are formed in such a culture can be picked off as described earlier. Chemical methods are sometimes 'messy', as they require the use of toxic or unpleasant-smelling compounds, eg. δ-camphor. As the maceration technique is so very simple and the results obtained almost instantaneous chemical means need only be used when dealing with species possessing wide hyphae which break easily during maceration allowing their protoplasmic contents to leak out.

IV. CHARACTERISTICS OF AGARICS IN CULTURE

A. INTRODUCTION

For over a century after the publication of Fries' *Systema Mycologicum*, the starting point for the nomenclature and taxonomy of the larger fungi, the agarics had been classified by relying solely on the morphological characters of their mature basidiomes. Fries scorned the use of even a lens, and it was not until the mid-1920s that the microscope was used to assist in the classification of agarics. It is true microscopy had been employed to great advantage by Fayod fifty years after the *Systema*, but it was generally only those mycologists studying moulds, mildews, etc. who regularly used the microscope. Similarly, culturing soon became part of the routine adopted by mycologists studying the ascomycetes.

Those studying basidiomycetes have largely ignored these advances. It should come as no surprise to learn that cultures of the vegetative phases of agarics in fact are not uniform and therefore are important in offering characters of very great potential in relating one agaric to another. Often associated with the vegetative phase are asexual propagules (diaspores), unknown until only fairly recently, or if known not appreciated as being part of the life-cycle of an agaric. Any additional characters which can be added to the few available at present for classification are always very welcome. Asexual stages are just such additional characters.

When fructification is induced in culture, the development of the basidiome can be followed and the reliability of the characters used by agaricologists in the field assessed. Unfortunately not all species have been induced to fruit in culture, but when basidiomes have been successfully obtained their study has been of very great significance.

Characters exhibited in culture have added enormously to the range of characters available in the routine work of the agaricologist.

B. DEVELOPMENT OF BASIDIOME PLATES 4-7

1. RANGE OF DEVELOPMENTAL TYPES

The agarics belong to the Homobasidiomycetidae and are characterised by their hymenium developing as a palisade of cells distributed over the surface of a hymenophore borne on what is correctly termed the basidiome (variously called carpophore, fructification, sporophore, basidiocarp or fruit-body). In the agarics the hymenophore is composed of plate-like structures called gills, and in the boletes the hymenium lines a spongy tissue composed of tubes. It is the hymenium which is composed of basidiferous cells; these cells, the basidia, bear the basidiospores.

Many agarics and boletes protect their developing hymenophoral tissue, either by the production of an all-embracing membrane or membranes, or by the growth of certain areas of the basidiome after the primordia have been initiated. Development can be studied by the examination of thin sections of basidiomes in different stages of development, and the differentiation of tissues and structures can be followed by utilising histological stains; eg., cresyl blue stain accentuates gelatinized tissues. These staining techniques are described on pages 53-57. From such studies a broad classification of the agarics and related fungi can be constructed. Dissection or the study of sectioned primordia with the aid of a low-power microscope unfortunately only shows gross differences, but even these allow one to demonstrate the major groups. It is now known that not all tissues are strictly comparable one species of agaric to another.

The following major divisions of basidiome development have been proposed:

i. Gymnocarpic (Gymno- naked; Carpous- fruit). Plate 5; Plate 6, fig. G.
When the hymenium develops to maturity on the surface of the basidiome or part of that basidiome, and is not enclosed in a cavity within the basidiome or protected by a veil, it is termed <u>gymnocarpic</u>. In this type of development, the pileus is formed from hyphae at the top of the stipe and subsequently expands by marginal growth. The hymenium forms later beneath the pileus peripherally and centrifugally. Some larger fungi, particularly some members of the boletes and their allies, commence gymnocarpic and only later become secondarily protected.

ii. Angiocarpic (Angio- hidden). Plate 6, fig. F.
When the hymenium appears, develops, and matures within a hymenial cavity which from the beginning is completely closed by a special envelope or veil, it is termed <u>angiocarpic</u>. This type of development characterises the heterogeneous group termed the gasteromycetes; this group appears to draw unrelated fungi together on this single developmental character. Some gasteromycetes are very closely related to the agarics and boletes, indeed some are more related to the agarics than they are to each other: see Appendix I.

iii. Hemi-angiocarpic (Hemi-helf). Plate 6, figs. A-E; Plate 7.
When the hymenium appears within a cavity but becomes exposed before it reaches maturity by the break-down of the enclosing membranes, it is termed <u>hemiangiocarpic</u>. This type of development characterises a very large number of our familiar agarics.

On sectioning young primordia formed in culture, or found in the field at the base of mature basidiomes, it can be demonstrated that this last developmental type, hemiangiocarpic, is rather complex, and several subdivisions can be recognized:

a. When there is an inner veil protecting the developing gills, it is termed paravelangiocarpic (para-near; velo-veil). Plate 7., fig. B.

b. When there is only an outer veil present, the development is termed monovelangiocarpic (mono- one). Plate 7, fig. A.

c. When there is no veil, but the pileus margin protects the developing hymenium, it is termed pilangiocarpic (pil- pileus). Plate 6, fig. D.

d. When there is no veil, but the pileus is seated in the stip-base which curls around to protect the developing hymenium, it is termed stipitangiocarpic (stipit- stipe). Plate 6, fig. C.

e. When the young pileus and stipe-apex are protected by an all-embracing tissue, the development is termed bulbangiocarpic (bulb- bulb, pertaining to the stip-base). Plate 6, fig. B.

f. When the pileus-margin is pressed tightly up against the stipe and the veil is obliterated, it is termed gymnangiocarpic (gymn- naked). Plate 7, fig. C.

g. When there are both outer and inner veils protecting the hymenium, it is termed bivelangiocarpic (bi- two). Plate 7, fig. D & E.

h. When hyphae from a whole range of tissues proliferate to grow and cover the developing hymenium, the development is termed metavelangiocarpic (meta- between, pertaining to the fact that this type of development is intermediate between the gymnocarpic and angiocarpic states; it is a modification of the gymnocarpic development and is a secondarily protected system). Plate 6, fig. A.

i. When hyphae from the margin of the pileus or from the basal regions of the stipe (or both) grow and cover the developing hymenium, it is termed mixangiocarpic (mix- mix, pertaining to the hyphae which grow out anastomose and interweave to form a protective membrane over the formerly exposed hymenium). This is a secondarily protected system derived from the gymnocarpic state. Plate 5, figs. B, C & E.

By careful sectioning three quite different conditions are found in the most complex developmental type, ie. bivelangiocarpic. These are termed schizohymenial, levhymenial, and rupthymenial and are distinguished as follows:

iv. Schizohymenial (schizo- splitting) Plate 7, fig. D.
When the gills are initiated within the tissues of the pileus and stipe and tear away from these tissues during maturation, they are termed Schizohymenial.

v. **Rupthymenial** (rupto- erupting).

vi. **Levhymenial** (levo- smooth) Plate 7, fig. E.
When the gills are initiated free from the stipe as a series of folds within a pre-formed gill-cavity, the development is termed **levhymenial**, and when initiated within the cavity but later become attached to the stipe, the development is termed **rupthymenial**.

The developmental types outlined above fall in the main into the families of agarics defined on morphological and anatomical characteristics. Thus although species formerly placed in the genus *Amanitopsis* apparently lack a ring, examination of the area at the base of the stipe shows a rudimentary annulus (ring) hidden by the volva. *Suillus granulatus* lacks a ring, but *S. albivelatus* possess a collar of white veil at the margin of the pileus which partially protects the developing hymenium at an early stage of development. In contrast, in *S. luteus* the margin grows so prolifically that it forms a sleeve around the stipe. The caulocystidia, which are initiated before the margin of the pileus extends, are still to be found under the sleeve and remain even after the veil splits at maturity to form the ring. This character can be more easily seen in *S. cothurnatus*, where the ring tends to stand out from the stipe. Thus, in keeping with the majority of boletes, these species of *Suillus* have a gymnocarpic development with a secondary veil-development superimposed onto the basic pattern.

Paxillus involutus has a basic gymnocarpic development which might have been predicted from its close relationship to the boletes in chemistry and anatomy. In this species the margin of the pileus itself protects the developing hymenium from the environment.

In addition to the results from the crossability tests described on pg. 93, the production of basidiomes in culture, followed by repeated studies of all stages of development, reveals useful differences between collections which may not have been apparent before or did not seem significant in the material when collected in the field.

Certain gasteromycetes undoubtedly have closer relationships to the agarics and boletes than they have with many other gasteromycetes. The majority of these fungi have been placed in the families Secotiaceae, Hymenogastraceae, Hydnangiaceae, and Asterogastraceae amongst the puff-balls and false truffles. The anatomy and micromorphology of members of the Secotiaceae not only run parallel with the families of the agarics, but several distinct interconnecting lines can be identified. Thus some members of the Secotiaceae have regular hymenophoral trama and pink, angular basidiospores, ie. Entolomataceae, others have coprinaceous characters, whilst others have bolbitiaceous characters, etc.

It would appear that these fungi have evolved in such a way as to take advantage of the primordial form; many gastroid basidiomes resemble enlarged 'undeveloped' primordia of more familiar genera. The gastroid forms are all united by lacking the ability to forceably disperse their basidiospores, and therefore failing to produce a spore-print when left humenium face-down on a glass slide of piece of paper. A few species with enclosed basidiomes are known to give some kind of a spore-print; these are classified as true agarics (Plate 14, fig. E). It has been even suggested that all the other secotioid and hymenogastroid fungi should be classified with the agarics.

Unfortunately, secotioid fungi have not fructified in any quantity in the laboratory, but nevertheless examination of primordia from the field, stained and sectioned in the way outlined in pgs. 52-54, allows the true relationships to become apparent. A table of examples of these connections between the gasteromycetes and agarics and boletes is given in Appendix I: see Plate 14.

Similar links between some of the basidiferous cup-fungi and the agarics can also be confirmed by thin sections stained as outlined earlier. With these staining schedules, the presence of the same tissues in the two superficially unrelated groups can be strikingly demonstrated. Once placed in a single family, even a single genus, the cyphelloid fungi can be easily shown to be a mixture of elements, and never so dramatically as by the use of histological stains.

When a lichen possessed circumscribed pits in the rind of the under-surface of the thallus, Acharius in the early nineteenth century termed the lichen cyphellate. The term became used generally when referring to these cup-shaped structures and then more specifically to all those fungi lacking asci whose basidiomes resembled these cups in overall appearance. Those fungi possessing disc-like basidiomes formed the basis of the family Cyphellaceae. Donk has shown how artificial this family really is; indeed the grouping has now been completely disbanded and its members placed in several, often quite unrelated, families. These transferences are based on similarities in anatomy and development between the cyphelloid fungi and members of other orders and families within the Basidiomycotina. A few examples of pleurotoid fungi, those with a lateral stipe or lacking a stipe, now take a natural intermediate position between the cyphelloid fungi and centrally stipitate agarics. All are similar to the agarics, which they resemble chemically, anatomically, etc. and in being gymnocarpic in their development. Examples are given in Appendix II: see Plate 15.

2. EXAMPLES

Angiocarpic Plate 6; Fig. F

Any gasteromycete, eg.*Lycoperdon* spp., *Scleroderma* spp.,

Geastrum spp. and most Secotiaceae see Appendix I. In culture *Cyathus stercoreus, Rhozopogon* spp..

Bivelangiocarpic Plate 7; Figs. D-F

Amanita spp., *Cortinarius* spp.. In axenic culture *Agrocybe cylindracea, Oudemansiella mucida, Pholiota terrestris, Hypholoma fasciculare, Psathyrella fimetaria, Stropharia semiglobata* and many *Coprinus* spp., eg.*C. cinereus, C.stercorarius* grp., *Psilocybe coprophila, P.cubensis & P. merdaria.* In dung culture (see pg. 109) *Coprinus cordisporus, C. ephemeroides, C. niveus.* In compost culture *Agaricus bitorquis, A. brunnescens & A. hortensis, Leucocoprinus birnbaumii.*

Bulbangiocarpic Plate 6; Fig. B

Volvariella spp.. In compost culture *Volvariella exculenta & V. volvacea.*

Bymnangiocarpic Plate 7; Fig. C

Conocybe pubescens & C. tenera groups. In culture *C. farinacea & C. pubescens.*

Gymnocarpic Plate 5

Any Cantharellaceae and Russulaceae; all Boletaceae lacking a veil. In axenic culture *Boletus amarellus, B. rubinellus, B. rubinus & Boletus badius; Lentinellus cochleatus, Schizophyllum commune, Pleurotus ostreatus* and other common pleurotoid fungi except *P. dryinus.* In compost culture *Clitocybe fimicola & Lepista nuda.* In culture with sterile trees *Laccaria laccata & Lacterius rufus.* Also *Polyporus ciliatus, P. brumalis* and many other true polypores. Also see: *Clitocybe tenuissima* (Bigelow H., Mycologia 62 (1970)), *Tylopilus felleus* (McLaughlin D., Mycologia 66 (1974)), *Calocybe indica* (Chandra, A., Mush. J. 40 (1976)).

Hemiangiocarpic

Any agaric which is not gymnocarpic

Levhymenial development Plate 7; Fig. E

Any annulate *Tricholoma* spp., eg. *T.terreum* complex

and *T. zelleri - aurantiacum* group, once placed in *Armillaria*.

Metavelangiocarpic Plate 6; Fig. A

    Any Gomphidaceae; In culture members of the *Armillaria mellea* complex.

Mixangiocarpic Plate 5; Fig. E

    In axenic culture *Lentinual edodes* & *Pleurotus cystidiosus* & *P. dryinus*. In culture with sterile trees *Suillus luteus*.

Monovelangiocarpic Plate 7; Fig. A

    Inocybe spp. & *Hebeloma* spp.. In axenic culture *Armillaria tabescens* & *Galerina mutabilis*. In culture with sterile trees *Inocybe petiginosa* & *Hebeloma populinum*.

Paravelangiocarpic Plate 7; Fig. B

    Flammulina velutipes; Bolbitiaceae (gymnangiocarpic development is a special case) *Agrocybe* spp., In culture *Agrocybe acericola* & *A. dura*. In dung culture (see pg. 103) *Coprinus bisporus* & *C. miser*.

Pilangiocarpic Plate 6;

    In culture *Paxillus involutus* & *P. panuoides*.

Schizohymenial development Plate 7; Fig. D

    Amanita spp.

Stipitoangiocarpic Plate 6; Fig. C

    Mycena stylobates group & *M. veneta*. As yet not cultured in the laboratory.

C. Secondary Spores

1. RANGE OF MITOSPORE MORPHOLOGY

    Most Fungi Imperfecti, excepting the Mycelia sterilia, are, or were once, thought to be connected only with ascomycetous perfect stages, and so mycologists concerned with the biology of fungi rarely directed their attention towards the basidiomycetes. There have been notable exceptions to this gap in our knowledge, although very few authors dealt with or even mentioned members of the Agaricales.

Several mycologists have been concerned with the nomenclatural problems arising when a species of fungus possesses more than one state, usually a perfect one (sexual) and an imperfect (asexual) state, although some species possess more than one of the latter. The necessity for special terminology was thought by the majority of mycologists to apply only to the ascomycetous fungi, but this is not the case, as many members of the Basidiomycotina, including the agarics, possess asexual or conidial states.

After much discussion, the state during which meiosis occurs has been termed the <u>teleomorphic state</u>, and other states exhibiting mitosis, eg. chlamydospores, conidia, spermatia, and rhizomorphs (see pg. 78) are termed the <u>anamorphic states</u> of that particular taxon.

<u>Conidial states</u> (Plates 8 & 9)

a. <u>Solitary Holoblastic Conidia</u> (Plate 9, fig. C.).

Such conidia are formed singly when a limited area of the wall of the conidiogenous cell or hypha becomes plasticised and blows out.

This is a small but nevertheless important category of conidia, as it relates the agaric genus *Hohenbuehelia* with the form genus *Nematoctonus*, a genus of nematode-trapping fungi classified in the 'Imperfect' Fungi and characterised by clamp-connections at the septa. Here the diaspore is formed at the apex of a long, thin hyphal branch resembling the sterigma of the basidium. Such conidia have been formed in cultures prepared from basidiospores; it has recently been demonstrated that basidiomes can also form in cultures of *Nematoctonus*.

Similar holoblastic conidia have been seen in *Pleurotus cystidiosus*, and in this species the conidia may be either mono- or dikaryotic depending on the genetic constitution of the hyphae which bear them; see page 78 for conidia-like bodies borne on hyphae in <u>Coprinus</u>.

b. <u>Thallic arthroconidia</u>: (Plate 8, figs. A-E).

Conidia are termed <u>arthroconidia</u> when they are formed by conversion and disarticulation of a pre-existing determinate hyphal element; if the element is a vegetative hypha they are described as <u>thallic</u>.

Undoubtedly diaspores of this type are the commonest conidia in the Agaricales. Although examples are found in nature, they are commonest in culture. They are probably one of the simplest ways of producing a conidium, ie. existing hyphae disarticulating at the septa.

Conidiophores at the one extreme are found on large, well-differentiated fruiting bodies, or at the other formed directly in clusters on the vegetative hyphae. The simplest types are the most widespread and are usually those seen most frequently in culture. Some of the arthroconidial stages have been given separate form-names when found in the absence of the basidiome; see examples on page 70.

The conidia form almost invariably on the vegetative mycelium by multiple septation and later separation of the units so formed. The conidiogenous hyphae develop either directly as branches of the main vegetative mycelium or as a group of laterals formed on these aforementioned branches. Certain gross correlations exist between families of agarics and particular conidial morphologies thus:

<u>Tricholomataceae</u>: conidia detached by fission of a double septa laid across the full width of the hyphae or formed by fracture.

<u>Cortinariaceae</u>, <u>Coprinaceae</u>, <u>Bolbitiaceae</u>, and <u>Strophariaceae</u>: conidia become detached by fracture or lysis of the walls of adjacent degenerating cells. Some slight thickening of the wall may take place, making it sometimes difficult to distinguish them from chlamydospores.

Depending on the genetic construction of the hyphae on which the conidia are produced, these asexual stages (ie. the conidia) may be either uninucleate or binucleate, and if the latter dikaryotic or monocaryotic, but usually dicaryolic.

Although formed mitotically, arthroconidia are not necessarily simple propagules. In many species they function as spermatia, not just by landing on a monocaryotic hypha and fusing with it but by producing chemical messengers. These chemical substances can be recognised by the hyphal tips in a colony, so that growth is stimulated in the direction of the condium. This has been called a homing response. Such responese apparently are not specific to a particular species but can occur between hyphae and conidia of closely related species. However, when this happens fusion between only slightly related taxa is usually followed by lethal reaction, ie. the protoplasm of the hyphal segments disorganises. Thallic arthroconidia are termed '<u>wet</u>' if coalescing into a droplet on the conidiophore and '<u>dry</u>' if the chaines of cells break up and do not coalesce. Apparently the wet conidia are highly surface-active and are spread in water perhaps because of their ciliate-like surface.

Arthroconidia of different species are on the whole morphologically similar to one another because of the lack of swelling, thickening, and ornamentation. However, by careful analysis general patterns become apparent; see examples on page 70. Thus in general, closely related agarics possess similar arthroconidia. Indeed members of a single section of a genus usually exhibit the same conidial morphology, whereas members of less closely related sections differ, often quite markedly. Morphology of arthroconidia can therefore suggest relationships. Sections which exhibit wide variation in conidial morphology reflect the possible un-naturalness of the group.

Agarics with arthrodonidia. The following two major categories of arthroconidia in the agarics exist:

(a) Arthroconidia are found directly on basidiomes of several members of the genus *Cystoderma*.
(b) Arthroconidia characterise the cultures of many species of *Coprinus* and *Psathyrella (Coprinaceae)*, many species of *Conocybe* (Bolbitiaceae), members of *Hypholoma* and *Stropharia* (Strophariaceae) and lignicolous members of the Cortinariaceae *(Gymnopilus & Pholiota)*, and *Mycena* spp. & *Flammulina velutipes* (Tricholomataceae). Arthroconidia in *Coprinus* seem to be wet and dry. Both wet and dry conidia can cause homing and may germinate. Wet conidia, however, are often poor at the latter and excel at the former; dry conidia are good at germinating but less often cause homing.

Members of the sect. *Setulosi* can be homothallic or heterothallic, bipolar or tetrapolar, and provided with or lacking clamp-connections. In addition, thallic arthroconidia are produced by some species, and these may be wet or dry depending on the taxon. For example:

In *C. congregatus* tufts of about 3-8 conidia are formed directly on normal-looking hyphae, whilst in *C. hiascens* and *C. sassii* short conidiophores are produced. In *C. pellucidus* and *C. stellatus* dry chains of conidia are formed.

In the *Coprinus micaceus-domesticus* group, conidia are formed in clumps of several conidia together, all of which readily germinate.

All species in section *Lanatuli* are tetrapolar and possess clamp-connections; their wet arthroconidia are formed in branched pin-heads held well above the agar surface.

c. <u>Chlamydospores</u> (Plate 8, figs. F-H; J-L).

Chlamydospores are resistant cells which are released by mechanical fracture of an undifferentiated cell-wall. These asexual spores are borne either terminally or intercalarily, and may be formed solitary or in groups.

These cells are thick-walled and are formed during the modification of a pre-existing cell by production of a new, thicker wall or by thickening of the pre-existing wall and the condensation of the protoplasm; often the thickening of the wall is accompanied by swelling, and retraction septa are frequently found.

(a) Solitary chlamydospores. Chlamydospores formed solitary are rather uncommon diaspores and are usually associated with other types of conidial production. They are found in several wood-rotting fungi, eg. *Pholiota aurivella*.

Chlamydospores with similar characteristics to gloeocystidia have been found in cultures of *Lentinellus* spp.

(b) Terminal and Intercalary Chlamydospores (Plate 8, figs. F-H) Agarics with this kind of characteristic conidial production include both widespread members of the genus *Nyctalis*. Here the conidial states are found independently in the field and have been given separate names (*Asterophora*; see pg. 71). The chlamydospores of the *Nyctalis* type exhibit a characteristic space between each conidium which is probably a result of condensation of retraction of the cell-contents; not all cells in a hypha necessarily produce chlamydospores. These structures are not uncommon in cultures especially of lignicolous agarics; eg. thick-walled cells in chains with a clamp-connection separating each unit characterise cultures of *Pleurotus dryinus*.

In the *Boletus sulphureus* group, chlamydospores are formed liberally in culture and are tawny or bright yellow; the individual conidia are usually separated from one another by at least one undifferentiated cell. In some agarics, eg. *Lyophyllum* spp., they may be separated by two or even more cells. The chlamydospores in *Boletus sulphureus* resemble the powdery yellow cells which often adorn the surface of the pileus in that species. Such cells bear a certain superficial similarity to allocysts which are found in cultures of species of *Pholiota* (see pg. 78).

Chlamydospores either form as the culture ages or are an integral part of the culture even in early stages. In *Coprinus trisporus*, chlamydospores form readily in culture, but they become pigmented as the culture ages. They form rows or bunches of cells superficially resembling the external cells of certain sclerotia. Smaller aggregates of cells are found in *C. cinereus* and are called microsclerotia. The aggregates of cells classified as *Attamyces bromatificus* - a stage of a *Leucoagaricus* sp. associated with the nests of leaf-cutting ants. Bromatium are formed from the aggregation of rounded swollen ends of the hyphae of the fungus (-i) cultivated and used by attine ants for food. When aggregated together they resemble *Aegerita*, of which one species, *A. duthei* is believed to be associated with the termite fungus *Raja (Termitomyces) eurhiza*.

    d.   <u>Modified clamp-connections</u> (Plate 9, figs. D-I & X)

This category is artificial in that it draws together unrelated groups of chlamydospores and thallic arthroconidia. Thus the arthroconidia in the *Sclerostilbum* stage of *Collybia racemosa* are joined to each other by clamp-connections, giving the unusual shape to the individual propagules. A similar explanation applies to the conidia of *Nothoclavulina* and *Antromycopsis*. Monocaryotic cultures produced from spores of *Pleurotus cystidiosus* are similar to those produced from macerated basidiome tissue. Both are classified as *Antromycopsis*, but whilst the former possesses thallic arthroconidia lacking clamp-connections, the dicaryon exhibits one at nearly every septum. In the closely related *Pleurotus dryinus*, chains of thick-walled cells, each joined by a clamp-connection, are formed in culture.

In *Squamanita* chlamydospores are formed sympodially, each subtended by a clamp-connection; these structures occur on the protocarpic tubers associated with the base of the basidiomes. Thus *Squamanita (Dissoderma) paradoxa* has smooth, irregularly shaped, thick-walled diaspores, whilst those of *S. pearsonii* are subglobose and possess very complex walls with internal structure (Plate 8, figs. J-L).

    e.   <u>Examples</u>

Thallic arthroconidia   Plate 8; Figs. A-E

    a.   Specialised fruit bodies (conidiomata)

Anamorph	Teleomorph
Nothoclavulina ditopa	*Armillariella (Arthrosporella) ditopa*
Sclerostilbum septentrionale	*Collybia racemosa*
Antromycopsis broussonetiae	*Pleurotus cystidiosus*

    b.   Found on basidiomes

Scattered on the pileus	*Cystoderma* aff. *longisporum*
Scattered on the stipe	*C. tricholomoides*

c. Developing in culture
  Bolbitiaceae   *Agrocybe dura*, *A. praecox* & *A. semiorbicularis* grp. Plate 8
    Conocybe pubescens grp. & *C. tenera* grp. Plate 8; Fig I.
  Coprinaceae   see below
  Cortinariaceae   *Gymnopilus penetrans*, *Pholiota* spp., eg. *P. terrestris P. squarrosa* & *P. aurivella* (in the latter sympodially arranged).
  Strophariaceae   *Hypholoma capnoides* & *H. fasciculare*,
    Psilocybe cubensis, *P. semilanceata*, *Stropharia semiglobata*: Plate 8; Gif. D.
  Tricholomataceae   *Flammulina velutipes*.
    Mycena inclinata: Plate 8; Fig.C.

Morphology

  Conidia casket-shaped or shortly rectangular   *Coprinus* sect. *Lanatuli*, eg. *C. cinereus* & *C. radiatus*.
  Conidia elongate often up to ten times as long as wide and slightly curved   *Coprinus* subsect. *Domestici*, eg. *C. domesticus* & *C. ellisii*
  Wet conidia   *C. cinereus*, *C. radiatus*, & *C. congregatus*
  Dry conidia   *C. pellucidus* & *C. stellatus*.

Chlamydospores

  Aggregated chlamydospores   *Coprinus trisporus*
  Solitary chlamydospores   *Pholiota aurivella*: Plate 9; Fig.B
  Gloeocystidia-like chlamydospores   *Lentinellus* spp.
  Tightly grouped chlamydospores   *Attamyces bromatificus* - *Leucoagaricus* Plate 9; Fig.A.
    Aegerita duthei - *Termitomyces*
  Microsclerotia   *Coprinus cinereus*
  Terminal and intercalary chlamydospores
   Anamorph   Teleomorph
    Asterophora lycoperdoides -roughened conidia   *Nyctalis asterophora*: Plate 8; Fig. G.

       Asterophora parasitica   *N. parasitica* : Plate 8; Fig. H.
         -smooth conidia   *Lentinellus* spp.
                                             Lyophyllum decastes: Plate 8; Fig. F.
                                           Pleurotus dryinus
                                           Boletus sulphureus grp.

Modified clamp-connections

    Anamorph   Teleomorph
        Sclerostilbum   *Collybia racemosa*, see pg. 70: Plate 9; Figs. D & E.

Modified clamp-connections

    Anamorph   Teleomorph
        Nothoclavulina   *Armillariella ditopa*: Plate 9; Figs. F & G.
        Antromycopsis   *Pleurotus cystidiosus:* Plate 9; Figs. H & I.
                                           Squamanita odorata complex: Plate 8; Figs. J & L.
                                           Dissoderma paradoxa: Plate 8; Fig. K.

Solitary Holoblastic conidia

    Anamorph   Teleomorph
        Nematoctonus   *Hohenbuehelia* spp.: Plate 9; Fig. C.

D. ANATOMICAL AND MORPHOLOGICAL CHARACTERS OF THE VEGETATIVE HYPHAE

1. CATEGORIES OF FEATURES

a. Bulbiloid bodies

The term "bulbil" has been misused for many years; it is now restricted to the complex structures produced by certain resupinate fungi. Those structures formerly called bulbils and for which names have not been coined are conveniently brought together as bulbil-resembling (ie. bulbiloid) bodies.

In *Mycena citriicolor*, a propagule often called a gemma is formed in place of the basidial tissue. This state has been called *Stilbum flavidum*. It is highly differentiated and resembles a reduced sterile dehiscent pileus hinged at the top of the stipe. It is now called a <u>cephalosus</u>. (Plate 9; fig. P-R.)

Coprinus clastophyllus under certain conditions forms pockets of sclerotised cells within the gill-tissue; such structures are termed <u>catervae</u>. A full spectrum of forms of fruit-bodies is known, ranging from a typical fertile, deliquescent coprinoid basidiome to an entirely asexual fruit-body composed entirely of catervae; intermediates possess varying numbers of groups of sclerotised cells and basidia. The pileus of the completely sterile fruit-body is pink in colour and has been called *Rhacophyllus lilacinus* (Plate 9, figs. J-O).

b. <u>Sclerotia and related structures</u>

True sclerotia develop from an aggregation of hyphae and have a centre of thin-walled cells. The outer cells are thickened and often pigmented, forming an outer rind. Such structures are found in a whole range of agarics.

Sclerotia may be the size of a pea or small walnut, eg. *Agrocybe arvalis*, or the size of an apple-pip, eg. *Collybia tuberosa*, or even smaller, eg. *Paxillus involutus*. In *Pleurotus tuber-regium* the sclerotium is very large and may be up to the size of a volly ball. When hyphae bind sand-grains or similar particles into a mixture of organic material, they form what is termed a <u>pseudosclerotium</u>, as in eg. *Panus badius;* this anamorphic state is called *Scleroma*.

Except in those species with true sclerotia, the sclerotised mass found in several agarics resembles an undeveloped primordium or a primordium with a redirected development. Sclerotised masses are composed of a variety of cells and are not arranged in a well-ordered manner as in a true sclerotium.

At the first stage of development, a hyphal lattice is formed by branching, anastomosing and re-branching of the hyphae. Numerous interconnections form between the intercalary hyphal cells present, and the lattices so produced become clothed with abundant aerial hyphae. A rounded aggregation of tightly woven hyphae is gradually formed. This aggregation subsequently develops into either the primordium or a micro-sclerotium. At the earliest stages it is impossible to say whether the hyphal aggregate will be one or the other. Even at the slightly later stages there is a striking similarity between the two. It has been suggested that in fact the primordia and micro-sclerotia are alternatives of the same developmental pathway and are really dependent on environmental conditions or stimuli (eg. light, temperature, and availability of carbon and nitrogen supplies) to launch them in one direction or the other.

c. Rhizomorphs and similar structures

In many agarics, the rhizoids or rhizomorphic growth exhibited are often a very impressive part of the life-cycle. Probably the most familiar are members of the form-genus *Rhizomorpha* generally considered a vegetative condition of *Armillaria mellea*. They are commonly called 'boot-laces' and are found in many members of the genus. Rhizomorphs can be produced in culture by growing *A. mellea* on almost any kind of medium.

The orange-brown vegetative growth found on wood and placed in *Ozonium auricomum* represents a state of several species of *Coprinus*, apparently all referrable to the subsections *Micacei* and *Domestici*. Such brightly coloured mycelial growth will develop from multi-spore isolates in the laboratory on 2% male extract agar or any other similar media.

d. Sterile fruitings

Abnormal fruiting bodies are not infrequently found in culture, perhaps because of a misbalance in nutrients, lack of light, or the atmosphere above the culture being too dry or stagnant. Such forms when found in nature have been described as independent species, often as members of the form-genus *Digitellus*; they are clavaria-like and produce sterile, finger-like growths. Such structures have been found in cultures of many wood-rotting agarics, particularly pleurotoid forms, (eg. *Lentinus lepideus*), and are frequent in the white-spored polypores, eg. *Polyporus squamosus*.

e. Additional cultural characters: Plate 10

By carefully standardizing the conditions for growth, the use of cultures has been successfully incorporated into the identification of the higher fungi, especially by those mycologists studying the wood-rotting members of the Polyporales and Agaricales. For these members, keys have been prepared not only to genera but to species based on their phenol oxidase activity, on characters of the clamp-connections, and on colony texture. Many lignicolous fungi possess laccase and/or tyrosinase, others neither; by incorporating gallic or tannic acid into the media it is possible to demonstrate their presence visually.

In comparative cultural studies, a rigorous reproducable procedure must always be adopted. The following is suggested: Grow the fungus for one week. From the colony so produced take an inoculum of actively growing hyphae and place it at the edge of a Petri dish containing about 30 ml.

gallic acid agar and incubate the dish for six weeks.

Although the above procedure is widely used in distinguishing commercially important fungi, when expanded to utilise other substrates it gives results which have proved to be of great potential in a better understanding of the species concept in Agaricales. Thus by replacing the gallic-acid agar by 2% malt of oatmeal agar, this technique has been found very useful in comparing the growth of terrestrial fungi. Indeed certain genera (eg. *Coprinus, Conocybe, Psilocybe,* and *Panaeolus*) (and in some cases species) can be immediately recognised by their vegetative growth on such media.

Taylor (1974 and 1977) has grown members of the Polyporales successfully on Basidiomycete agar (see pg. 17) to which instant skimmed milk has been incorporated. The agar is prepared by making a paste of 2 g of skimmed-milk powder and adding it to 100 ml of agar at $55^{\circ}C$. Preliminary results indicate that differences between colonies of closely related species can be accentuated by adopting this technique.

The rapidity of growth of the culture, the colours of the culture (the mat) both when young and when mature, and the colour of the medium below the cultures are often indicative of a group of closely related species or even of a genus. The amount of aerial hyphae and of submerged hyphae is also of importance, as is whether the culture emits light when examined in the dark (pg. 79).

Usually, a culture exhibits in time a particular set of macroscopic characters which may help to indicate relationships between it and other cultures, or reveal anomalous species within a genus or family. However, of greater importance in such an assessment appears to be the presence or absence of certain microscopic characters. Such characters are fundamental to the identification of larger fungi in culture including agarics and are listed in the following paragraphs.

Branching pattern (Plate 10, fig. V)

Mycelial growth may differ markedly in cultural morphology between species. The branching may be either at acute angles from the main branch or nearly at right angles to it. The distances between successive branches also differ from species to species; they may be very long, or the whole pattern may be considerably telescoped. Although such features do not allow a specific identification to be made, they are constant for a species and are often correlated with other characters.

There is usually a difference between monokaryotic and dicaryotic hyphae, not only in the characteristics of the branching, but also in the size of the colony produced in a given period or the extent of the development of aerial or submerged hyphae.

Most hyphae of ascomycetous fungi are usually broad when compared with those of the basidiomycetes, although there are some exceptions to this general statement, eg. submerged hyphae of *Coprinus narcoticus* group.

Bulbils as outlined by forest pathologists in their widespread keys on identification of cultures are small knots or compact masses of hyphae where branches have twisted around a central hypha; see Plate 10, fig. C. They resemble multiple clamp-connections (see below). They must not be confused with the propagules formerly called bulbils and mentioned on page 72.

In some cultures, the hyphae branch at right angles and have refractive walls, or the hyphae form clumps of repeatedly branched usually dichotomous side branches. The latter are called stag-horn hyphae.

In yet other cultures the hyphae swell often to 25μm or more in diameter and become deeply stained by solutions of phloxine, of trypan blue or of cotton blue. (see pgs. 15 & 16). Such hyphae are termed laticiferous hyphae. The swellings may be widespread or restricted to a single cell. They parallel similar hyphal modifications found in the basidiome.

One type of branching is unique to the vegetative hyphae of the basidiomycete, although it is not always present even in this group of fungi. It is characterised by the backward growth of a short hyphal branch from a site in front of a septum and its fusion with the cell immediately behind that septum. These structures are called clamp-connections.

Clamp-connections

Much emphasis has been placed in the past on the presence or absence of clamp-connections in a culture or in the basidiome tissue. The presence of clamp-connections in a culture under standarised conditions is always a reliable character; the ability to form clamp-connections has therefore been used in the classification of cultures of the larger fungi. In some experiments the fungus has never been taken to its ultimate conclusion (i.e. basidium production), the presence of clamp-connections being taken as sufficient evidence to indicate complete compatibility. Care must be taken, however, to ensure that the conditions are standardized, as it has been demonstrated in several taxa that the production of clamp-connections is dependent on the medium used, and in some cases they may be totally absent in culture although present in the original basidiome. Many active field mycologists know that in a single fruiting population the number of clamp-connections varies from one part of the basidiome to another and from vegetative to spore-producing tissue.

Clamp-connections when present are found either consistently throughout the culture or only on the hyphae of the oldest parts of that culture. In many boletes (eg. *Suillus* spp.) paarige branching is found; this is a complex multiple branching from a single clamp-connection: Plate 10, fig. N. Some non-agaricoid fungi possess whorles of clamp-connections at a single septum, and these are termed <u>multiple</u> clamp-connections. They resemble to some degree paarige branching.

The shape and position of the clamp-connection is helpful in identification. Whereas some clamp-connections are very prominent and have a central perforation, other clamp-connections are rounded, and yet others have an overall, rather rectangular shape. '<u>False clamps</u>' are those structures formed when a side branch commences to grow backwards in an attempt to form a clamp-connection but fails. (Plate 10, fig. V)

Although a species will possess clamp-connections, in the dicaryotic phase all monocaryotic hyphae lack clamp-connections. Hyphae possessing these structures are often said to be <u>nodose-septale</u>. The French call these same structures '<u>bouclé</u>', and an abbreviation found in many texts is simply '<u>clamp</u>.'

<u>Tissue types</u> Plate 10, figs. Q, S and U.

In many cultures the hyphae aggregate into distinct tissues. Thus round or ovoid, thin-walled cells with intercellular spaces typify species of some agaric genera in culture, whereas species of other genera exhibit polyhedral, often pigmented, thin- or thick-walled cells without intercellular spaces. Some other genera possess inter-woven hyphae with spaces between hyaline usually thin-walled cells, and yet others in culture possess tough membranaceous tissue adhering tightly together but usually with interhyphal spaces. For example, *Armillaria mellea* in culture exhibits interwoven or parellel thick-walled pigmented cells and specialised structures called rhizomorphs. (See pg. 74).

Hyphae of some species of larger fungi produce numerous interlocking projections which form a pseudoparenchymatous layer. <u>Cuticular cells</u> are swollen cells whose contents, although hyaline, stain strongly in solutions containing phloxine or cotton blue; walls of these cells are unstained with dye and either remain colourless or become brown with maturity.

Hyphae with thick, refractive walls and narrow or apparently lacking lumina are observed in many species of polypores. They are found in the basal parts of basidiomes of some pleurotoid fungi and may also be seen in cultures of these same agarics. Such hyphae are usually hyaline, although they may become coloured with age and in very old herbarium specimens. Comparison should be made with the mitic system of hyphae utilised in the classification of the polyporaceous fungi. (Corner, 1932).

Allocysts and related structures

End-cells and penultimate hyphae are differentiated in cultures in many species, some having the end of the hyphae coiled or spiralled, ie. helicoid, or exhibiting tree-like projections, ie. dendroid. Others become covered in encrusted material, frequently considered calcium oxalate, but in some species undoubtedly of mucopolysaccharide origin: Plate 10, figs. E-I.

Allocysts are chlamydospore-like structures found in *Pholiota gummosa*, *P. alnicola* & *P. conissans*, and *Coprinus radians*. They are often terminal and do not appear to be functional propagules; they act perhaps as repositories for unwanted by-products of metabolism: Plate 10, fig. R.

Cystidia are differentiated, hyaline structures produced on aerial mycelia and resembling sterile cells of the same name found in and on the basidiome. Those on the basidiome have been used extensively in the classification of the agarics and are described in Book III of this series. When these sterile cells are yellow or brown, thick-walled, lanceolate, and darkened by solutions of alkalis, they are called setae. Setules are similarly shaped structures but are thin-walled or are only slightly thickened towards their bases.

Gloeocystidia are cystidia which have oily contents and stain strongly with cotton blue; they also stain with sulphoaldehyde and similar reagents as outlined in Book III.

Refractive bodies

Amorphous terminal or intercalary, yellow refractive bodies are found in many different genera. They can be accentuated if stained in phloxine, or erthrosin in ammoniacal solutions (pg. 15), or viewed with a phase contrast microscope. Their role is unknown. The production of crystals both in the agar medium and attached to the individual hyphae is often characteristic of certain species of agaric. Their formation may be more significant than at first thought, and their appearance should be monitored more carefully in the future.

Diverticulae

The two members of *Coprinus* sect. *Comati* so far studied in culture produce on the mycelium small spherical often quadrate bodies, resembling candelabra, which neither germinate nor attract hyphae of the same or different species. Their function remains a mystery. *Schizophyllum commune* produces in culture small wall-elongations on the hyphae. These diverticulae often have small apical prolongations, and they have been confused with conidia.

f. <u>Bioluminescence</u>

Cultures of certain groups of species when examined in the dark show a marked phosphorescent glow--a blue-green light which lights up part of the entire culture-vessel. *Armillaria mellea* and its close relatives are particularly good in exhibiting this characteristic. Other species are *Panellus stipticus* (but only North American strains), *Mycena citriicolor* (=*Omphalina flavida*), and *Omphalotus olearius* (Jack O'Lantern).

In cultures maintained at 21-25°C, luminescence will occur after three weeks or so and may last for a considerable time. The culture must be aerated, as oxygen is required for the reaction to take place. One must have patience, as the light emitted can only be seen after a 10^+ minutes stay in a darkened room; this period will allow the eyes to become accustomed to the environment. Best results are obtained when the hyphae are grown on bread-crumb agar prepared as outlined on pg. 6. Indeed several other agarics (eg. *Mycena galopus* and *M.polygramma*) also become luminescent on this medium, a phenonemon which otherwise might be overlooked.

Fresh basidiomes of *Panellus stipticus* will produce viable basidiospores for a week or more if kept moist, even if the original collection had been quite dry when found. Germinate the spores shed from a fresh basidiome on 2% malt extract agar or similar medium and allow them to grow for two or three weeks at room temperature, around 22°C. A white fluffy mycelium will result, which will give positive results if viewed in the dark or allowed to come into contact with photosensitised paper. Transfer an actively growing part of the culture to bread-crumb agar and leave for a few days. Then place a piece of panchromatic film or paper, emulsion side down, over the Petri dish and leave for varying lengths of time from 4-24 hours in a dark room or blackened box. Develop the film in the normal way, when a shadowy image will be revealed. The method can be modified by cutting the shape of a letter out of a piece of card, mounting a piece of film over the cut-out, and then placing this on the Petri dish containing the fungus under study. On developing the film the outline of the cut-out appears.

Panellus stipticus will fruit in pure culture if grown on paper-pulp medium, saw-dust medium, maize-kernel medium, or bread-crumb medium, all prepared as outlined on pgs. 6 & 10. The basidiomes so produced are also bioluminescent, the strongest light being emitted from the gill-surfaces and stipe. Light is emitted best at room temperatures, although radiation can be registered from below 5°C to over 30°C. Those wishing to record the radiation by light-meter can do so easily as the light is in the band 470 - 640 mµ; a commercially available photographer's meter is adequate.

g. <u>Examples</u>

Allocysts   *Pholiota alnicola* & *P.gummosa*;

Coprinus radians

Caterva
    Anamorph   Teleomorph
        Rhacophyllus lilacinus   *Coprinus clastophyllus:* Plate 9.
                                                        Figs. J-M.

Cephalosus
    Anamorph   Teleomorph
        Stilbum flavidum   *Mycena citriicolor:* Plate 9.
                                                        Figs. P, Q and R.

Conidiophores
    Diversity--in *Coprinus* subsect.
        Setulosi
            Short conidiophores   *Coprinus sassii* & *C. hiascens.*
            Conidiophores reduced to peg-
                like projections   *Coprinus congregatus.*

Candelabra cells   *Coprinus comatus.*

Diverticulae on hyphae   *Schizophyllum commune:* Plate 10.
                                                        Fig. A

Diverticulae on velar cells   *Coprinus stercorarius* grp.

Ozonium
    Anamorph   Teleomorph
        Ozonium auricomum   *Coprinus domesticus* complex.

Pseudosclerotium   *Panus badius* (anamorph sometimes
                                                        called *Scleroma*).

Rhizomorph
    Anamorph   Teleomorph
        Rhizomorpha subcorticalis   *Armillaria mellea*--white sheets
                                                        beneath bark.
        Rhizomorpha subterranea   *A. mellea*--boot-laces.
        Rhizostroma xylostroma   *Tricholomopsis(Megacollybia)*
                                                        platyphylla

Sclerotium see also *Aegerita duthei* pg. 70
     Agrocybe arvalis, Bolbitiaceae.
     Coprinus sclerotiger & *C. tuberosus*, Coprinaceae.
          microsclerotia--*C. cinereus.*
     Collybia cookei & *C. tuberosa* Tricholomataceae.

Lepiota meleagris (*Coccobotrys xylophilus* as anamorph) &
   Leucocoprinus birnbaumii, Lepiotaceae.
Paxillus involutus in cultures with sterile trees,
   Paxillaceae.
Pleurotus tuber-regium, Pleurotaceae.

Sterile fruitings
   Lentinellus cochleatus & *L. castoreus*.
   Lentinus lepideus (in dark culture) also *Polyporus squamosus*.
   Compare with gastroid fruitings of *Lentodium squamulosum* in
      populations of *Lentinus tigrinus*, Pleurotaceae: Rosinski
      & Robinson <u>Amer. J. Bot.</u> 55; (1968).

2. CULTURE FORMULAE

   a. <u>Criteria</u>

   A short-hand method of describing fungal growth in culture is now employed by a large number of mycologists, particularly forest pathologists. It was designed primarily for use by support staff to allow fairly rapid identification of cultures of lignicolous fungi. The coding prepared for each species is built up from observations on the characters of several isolates taken from basidiomes of a chosen species. The key pattern is formulated by listing the diagnostic characters of the species in the order set out below and choosing from each section the correct digit. Those which have variable characters can be represented by more than one digit; those characters which do not vary occur only once in the formulae. Thus provision is made for all possible variation.

   A. 1. On broad-leaved wood
       2. On coniferous wood.

   B. 1. Mat remaining white, pale yellow, or pale pink, for six weeks.
       2. Mat yellow or brown, at least when mature.

   C. 1. Decolorises medium in gallic and tannic-acid agars: see pg. 83.
       2. Diffusion zone lacking.

   D. - E. For explanation of characters see pages 72-79.

   D. 1. Clamp-connections present throughout cultural mat.
       2. Simple septa present only.
       3. Clamp-connection occuring only in old parts of culture.
       4. Multiple clamp-connections present.

   E. 0. Contorted or incrusted hyphae present.

        1. Cystidia or gleocystidia present. Plate 10. Fig. K.
        2. Setae or setal hyphae present. Plate 10. Fig. L.
        3. Knots of hyphae ('bulbils') present: see pg. 72. Plate 10. Fig. C.
        4. Rigid hyphae with right-angled branches present.
        5. Cuticular cells forming psuedoparenchymatous layer present.
        6. Hyphae present with numerous interlocking projections. Plate 10, Fig. O.
        7. Swelling on hyphae present.
        8. Laticiferous hyphae present. Plate 10. Fig. P.
        9. No special structures present.

F. 1. Chlamydospores present: see pg. 68. Plate 8. Fig. F-H.
    2. Chlamydospores apparently not formed.

G. 1. Holoblastic conidia produced: see pg. 66. Plate 9. Fig. C.
    2. Conidia apparently not formed.

H. 1. Thallic arthroconidia present: see pg. 66. Plate 8. Figs. A-E.
    2. Thallic arthroconidia not formed.

I.   1. Rate of growth rapid (plates covered in one or two weeks).
    2. Rate of growth moderately rapid (plates covered in three or four weeks).
    3. Rate of growth slow (plates covered in five to six weeks).

J. 1. Fruiting before the end of six weeks.
    2. No fruiting observed.

K. 1. Reverse of culture brown, at least in part, before the end of six weeks.
    2. Reverse unchanged, or not darker than honey-yellow, in six weeks.
    3. Reverse bleached, at least in part, before the end of six weeks.

b. <u>Examples of formulae</u>

Armillaria mellea

    This agaric occurs on coniferous and broad-leaved trees. The cultural mat is lilaceous brown and actively decolourises gallic-acid agar. The cultures lack clamp-connections, chlamydospores, conidia, and spermatia, but a pellicle of dark cuticular cells is produced; a dark-coloured reverse to the culture is also formed. The rate of growth is very slow compared with many other basidiomycetes, and fructifications are very rarely obtained except after special treatment. The formulae would therefore be:

Armillaria mellea   (1 2) 2 1 2 5 2 2 2 4 2 (1 2)

A. mellea is undoubtedly a mixed taxon, and there is every likelihood that more research will confirm the presence of two (if not more) distinct taxa suggested by the double entries for items a and k.

Some other common lignicolous agarics are characterised by the following formulae:

Flammulina velutipes	1 2 2 1 (8 9) 2 2 1 2 2 2
Lentinus lepideus	2 1 (1 2) 3 9 1 2 2 2 2 2
Lentinus tigrinus	1 (1 2) 1 1 9 1 2 2 2 2 2
Lyophyllum ulmarium	1 1 2 1 9 2 2 2 (3 4) 2 2
Oudemansiella radicata	(1 2)(1 2) 1 1 (6 9) 2 2 2 2 2 2
Pholiota adiposa	(1 2)(1 2) 1 1 9 2 1 2 (2 3) 2 2
Pleurotus ostreatus	(1 2) 1 (1 2) 1 9 2 2 2 2 (1 2) 2
Schizophyllum commune	1 1 (1 2) 1 9 (1 2) 2 2 2 (1 2) 2
Xeromphalina campanella	2 2 1 1 6 2 2 2 4 2 2

When the key formulae are arranged in numerical order, it becomes apparent that closely related species are often parallel or identical in their digit pattern. In the latter group, other characters of the culture such as texture of the mat, colour, ie. those characters not incorporated into the formula, must be used to separate species.

It must be emphasised that the culture-formulae are applicable only if the characters are determined under precisely the same prescribed conditions.

c. Procedure

Grow the isolate on a Petri dish for one week on 2% malt agar prepared as outlined on pg. 3. From an actively growing culture cut out cubes two or three millimetres square and transfer to the edge of each of five 90mm Petri dishes containing about 20 ml of malt agar. Incubate the cultures at room temperature in the dark, exposing to the light only for examination. (When cutting out small pieces of the culture for examination make sure that the scalpel has been flamed.) Describe all the macro- and microscopic characters once a week, every week for six weeks. Reference to all of the five cultures allows characters to be recorded with indecision.

Petri dishes containing gallic- or tannic-acid agar prepared as outlined on pg. 4 should be inoculated in the centre. The inoculum should be four or five millimetres square and should be taken from actively growing cultures four to six weeks old. A positive reaction is registered when a colourless zone is formed about the inoculum; activity can vary from weak to strong, or may be completely absent (see below).

For more sophisticated techniques, the references at the end will be of assistance. Reference to J.Stalpers key to wood-rotting fungi in *Studies in Mycology* No. 16 (Baarn): 1978 will show the excellence of this approach.

V. EXPERIMENTS, STUDIES AND TESTS

A. CHEMICAL TESTS

1. GALLIC-ACID TEST

Nobles (1948) utilised the gallic-acid test to determine the identity of lignicolous agarics and polypores. The gallic-acid test is based on the production by the agaric of organic catalysts (enzymes) called extra-cellular oxidases; these particular enzymes assist the chemical dislocation of lignin-like molecules in the substrate. The gallic-acid agar is prepared as outlined in the recipes on pg. 4 (but see next paragraph). Nobles' approach is an essential first step towards meaningful assessment of the characters of a culture; in addition it allows the relationships with other genera or species to be ascertained when only a culture is available.

After growth on the gallic-acid medium for 6 weeks, two major groups of cultures can be detected a) those cultures where a clear zone is produced around the culture, and b) those cultures where apparently no reaction is to be found. The clear zone can be accentuated if the gallic-acid agar is poured, when almost ready to set, onto a base of tap-water agar already in a Petri dish. Some species, eg. *Schizophyllum commune,* are assignable to both category a) and category b) because activity is present in some isolates and absent in others. This is unusual; the groupings indicated are usually constant. A few cultures, however, do not fit comfortably into either one category or the other, because the activity when present is rather weak.

Gum guaiac allows a visual assessment to be made. A solution of 0.5% of gum guaiac in 30 ml of 95% ethanol (prepared by mixing 95 ml ethanol and 5 ml water) is dropped onto the actively growing mycelium. If the enzymes are present, a blue colouration is produced: see pg. 87.

Phenol oxidase modifies a whole range of organic compounds, and the constituents of the malt agar on which fungi are grown in culture are no exception. On page 82 three categories are noted under key character K, but in fact, after growth in culture for six weeks, at least six groups can be recognised which can be correlated with the gallic-acid tests. The categories are:

    lacking—no brown colouration.
    very weak (+)—diffusion zone, light to dark brown, under
        inoculum, only at centre of mat and not visible from above.
    weak (++)—as above but zone under the entire mat.
    moderately strong (+++)—diffusion zone light to dark brown,
        extending a short distance beyond colony and therefore
        visible from above.
    strong (++++)—diffusion zone dark brown.

very strong (+++++)—diffusion zone very intense and forming a distinct wide corona around the mycelial mat.

2. BIOCHEMICAL TESTS

Taylor (1974; see Appendix) for the identification of cultures of basidiomycetes has introduced several biochemical tests similar to those used in routine bacteriology. These tests are of much wider application than the gallic/tannic acid test described above for lignicolous fungi, and have already been successfully applied to agarics.

Agarics are heterotrophic fungi, and therefore, in order to break down complex organic molecules they require an external supply of sugars, simple amino-acids, and often growth substances such as vitamins. In culture these compounds must be supplied by the medium. Enzymes have evolved in the fungi to bring about specific reactions, and many pass from the hyphae into the medium. By supplying suitable known substrates, the enzymes present can be detected by the products of the reaction or the destruction of the substrate, which ma be indicated by a colour change. Taylor's tests are based on these enzyme reactions.

Three groups of tests are described, each group depending on the period of time required for the reaction to be completed: instant, short-term, long-term incubation tests.

a. and b. Instant & Short-Term Incubation Tests

Procedure: Grow the agaric at $22°C$ for 3 days on Basidiomycete agar prepared as directed on page. 6. Using a sterilised cork borer, cut out two plugs 3mm in diameter from the edge of the colony. Transfer the plugs to the test medium, which has previously been placed on porcelain spot-test tiles or microscope slides with central depressions. In order to prevent the agar from drying out, the tiles or slides should be kept in a large Petri dish, a flat sandwich box, or the damp chamber described on pg.28. The test should be repeated at least three times to obtain reproducible results; also repeat with plugs from a 5-day old colony.

When pipetting the reagents in the tests below, always ensure a bulb-pipette is used. DO NOT suck up the reagents with a mouth pipette.

a. Instant Tests

1. Catechol: 25 mg of catechol in 100 ml of citrate buffer prepared as outlined on pg.12 is pipetted into the wells containing the plugs. If catechol-oxidase is present, a brown colouration or

darkening effect is produced within 6 hours.

2. <u>Peroxidase-cyanide stable</u>: A drop of freshly prepared 1% pyrogallol (1 g in 100 ml water) is pipetted onto the test plug, along with a drop of a 0.4% aqueous solution of hydrogen peroxide (40 mg of peroxide in 100 ml water). The production of a bright golden yellow colour within 30 mins indicates a positive reaction.

A parellel test using indoxylacetate has been designed for recognising the esterase-fluoride stable: see Taylor; Appendix IV.

3. <u>Gum guiac</u>: Dissolve 50 mg of powdered gum guaiac in 30 ml of 75% ethanol prepared by mixing 75 ml of ethanol and 25 ml water. A positive reaction is indicated by the production of a blue colouration when the solution is added to the plug.

4. <u>Lactophenol</u>: Stir together one part each of phenol and lactic acid, and add the mixture to the plug. A positive reaction is indicated by the production of a brown colour.

5. <u>Tyrosin</u>: Prepare a tyrosin suspension in 95% ethanol and pipette onto the plug. Darkening indicates the presence of the enzyme tyrosinase.

b. <u>Short-Term Incubation Tests</u>

6. <u>Cinchonidene</u>: 40 mg of cinchonidene is added to 100 ml of 1.2% water agar, prepared by adding 12 g agar-agar to 1000 ml distilled water as outlined on pg. 8, and the plugs placed on top. Clearing around the plugs after 24 hours indicates that activity has taken place.

7. <u>Ferulic acid</u>: 6 mg of ferulic acid is added to 15 ml of Basidiomycete agar prepared as outlined on pg. 6, and the plugs placed on the agar when set. After 24 hours, a red or pink pigmentation indicates an active culture.

8. <u>p-Nitrophenyl compounds</u>: 10 mg of o-nitrophenyl-fucoside is added to 1 ml of acetone and 9 ml of citrate buffer, and the whole pipetted onto the plugs in the wells. After 24 hr, a drop of 5% sodium bicarbonate (5% $Na_2Co_3.10H_2O$ in 100 ml water) is added to each, producing a rich yellow colour if p-nitrophenyl has been released by the enzyme (-fucosidase). The reaction can be repeated for a whole range of enzymes which break down simple sugars: see Taylor; Appendix IV (p. 127). Modified nitrophenyl tests can be conducted for phosphatase enzymes: see Taylor, 1974.

9. <u>Rutin</u>: 100 mg of rutin is added to 1 ml of methanol (Methyl

alcohol), and the whole diluted with 9 ml of 1.2% water agar prepared by adding 1.2 g agar-agar to 100 ml water and boiling. After 24 hrs, a dark colouration is produced when the culture exhibits activity.

10. <u>Starch</u>: 1% soluble starch buffered to pH 5.4 with buffer prepared as outlined on pg. 12 is pipetted onto the plug in the well. Leave for 24 hrs and then add a drop of Lugol's iodine made up as described on pg. 18. The presence of a brown colour indicates that the starch, which could have been blue in iodine, has been destroyed by the enzyme.

11. <u>Vanillin</u>: Vanillin powder is dispersed in 1.2% water agar prepared as in 3 above. Dark-coloured, spherical, non-cellular bodies are produced in and around the plugs within 24 hrs when the fungus is active.

c. <u>Longer-Term Incubation Tests</u>

<u>Procedure</u>: An inoculum is taken as in the tests above from a 5-day old culture grown at $22^{\circ}C$, mixed with the test substrates, and incubated for 1, 2, 3, and 4 days at $22^{\circ}C$.

12. <u>Blood</u>: 1 ml of human blood from which the serum has been removed is added to 9 ml of Basidiomycete agar just before boiling. A positive result is recorded if concentric zones of hydrolysis and precipitation are found around the fungus.

13. <u>Gelatine</u>: 15% aqueous (w/v) gelatine is melted and sterilised. Antibiotics prepared as outlined on pgs. 13 & 14 are added and the mixture poured into Petri dishes. The dishes should be refrigerated after pouring to cool rapidly before incubating with the culture. Liquefaction zones after 3 days indicate that a gelatinase enzyme is present.
    Alternatively, 40 mg of gelatine can be incorporated into water agar and the culture tested. Flooding the culture with an acidified solution of mercuric chloride, prepared by dissolving 1.5g $HgCl_2$ in 2ml conc. HCl and 10 ml of water. Unchanged gelatine does not clear.

14. <u>Lecithin</u>: 10% (v/v) of freshly prepared, centrifuged, unsterile lecithin supernatant is added to Basidiomycete agar and cooled overnight in a refrigerator before inoculating. Any clearing around the inoculum indicates a positive reaction and the presence of lecithinase.

15. **Lipid**: 1% tributyrin (v/v) is incorporated into the Basidiomycete agar before inoculating. Cleared zones extending beyond the inoculum are indicative of a positive reaction. They can usually be seen more distinctly if the test agar has been poured, when just about to set, onto a base of tap-water agar.

16. **Milk**: Grow each isolate for 2 weeks in a 90mm Petri dish of Basidiomycete medium incorporating 2% instant skimmed-milk powder (Taylor used New Zealand Co-op) (2 g of milk powder is made into a paste with 100 ml of the Basidiomycete agar at $55^{o}C$ and the whole added to 900 ml of the same agar). The mixture is adjusted to pH 5.0 with 0.35% N-hydrochloric acid, ie. 0.35 ml of that prepared on pg. 19 with the addition of 100 ml water. Clearing of the agar after 4 days indicates a positive reaction.

Alternatively, 2% instant skimmed milk powder can be added to 1.2% water agar (pH 5.5) prepared as in (3), at $50^{o}C$ and shaken gently. Leave for 20 mins in a water-bath, add 0.35% N-hydrochloric acid, and again shake gently before inoculating with the culture. If any cleared or opaque zones are formed around the isolates after 2 days, the test is considered positive.

3. ADDITIONAL REACTIONS—SPOT TESTS

In addition to the tests compiled by Taylor (1974 & 1977), other workers have used a whole range of aromatic compounds as test-substrates. Tests in these studies are conducted on 2% Malt agar in Petri dishes by inoculating with a culture 15 mm from the edge of the dish and allowing the mycelium to grow for 4 days at $22^{o}C$. The test solutions are prepared as 0.1 M solutions in 95% ethanol, prepared by mixing 95 ml of ethanol with 5 ml water. The solution when thoroughly mixed is pipetted onto the culture: but first see page 86. Thirty-two different substances have been listed by Kaarik (see refs. in Appendix IV), but as some of these are potentially carcinogenic great care must be taken. A current copy of the <u>Documentation of Threshold values</u> (1977) should be consulted; see Appendix IV.

A few chemicals which are recommended are:

gallic acid	tannin	phenol
resorcinol	quinol	thymol
naphthol	aniline	phloroglucinol
orcinol	p-quinone	p-cresol

It will be seen that gallic acid and tannin, two compounds used in Noble's tests, have also been included. This draws the two methods close together and allows comparisons to be made; see pg. 83.

4. BASIDIOME MATERIAL

Chemical tests have been widely used in the taxonomy of agarics; indeed with careful use and rigorous interpretation they have been utilised to suggest or substantiate possible relationships between species and between genera.

It has been customary in the past simply to rub or drop the chemical onto the flesh, gills, or surface of the basidiome. However, such treated tissues collapse on drying and cannot be revived, thus making the specimen unsuitable for the herbarium.

By capillary action the chemicals cover a larger area than intended and may draw together different chemicals. It is therefore suggested that with fleshy fungi the tissue be cut into small cubes 3 mm^2 and added to about twice their volume of reagent in a test tube, or cut into discs which can be tested in welled slides. These reactions are discussed further in 'Book I' of this series; reference to a full list of reagents and reactions is given in the reading list.

B. GENETIC STUDIES (PLATE 12)

1. INTRODUCTORY DISCUSSION

Genetics is that branch of biology dealing with the resemblances and differences between organisms related by descent, and the nature and distribution of those heritable pairs of factors called genes. The genes bring about a particular pattern of development in an off-spring and influence the production of a certain set of characters; each gene is carried on a thread of proteinaceous material, several threads making up each nucleus of each cell.

The gametes which carry the genes, be they egg-cells, pollen grains, etc., have a single complement of these genes, and only when the gametes have fused is the pair of genes reformed. In organisms such as man and his domesticated pets, flowering plants, etc. a single gene is offered from each parent, and by fusion of gametes new, slightly different genes are brought together in the off-spring. The gametes are called haploid and the fused fametes, or zygotes, are called diploid.

Agarics are placed in the group of fungi called the Basidiomycotina, and this group differs substantially not only from other groups of fungi but from other organisms. They differ in that the fusion of the gametes to form a zygote is delayed for a very long period, perhaps even years. Each vegetative cell of the basidiomycete possesses two nuclei, but although they are engulfed in the same cytoplasm they are quite separate and can operate independently, or in various degrees of mutual dependency. It is only in the basidiome (fruiting body) that these nuclei fuse, exchange genes, and produce spores which will propagate a new generation. Fascinatingly, it is only in the end-cell of the hymenium, ie. in the basidia, that this fusion takes place, and all other cells of the stipe, pileus and gills contain two independent nuclei. The cells are called <u>dicaryotic</u>. This phenomenon makes agarics particularly interesting organisms to study. When the spores formed by the reproductive division are dispersed and begin to germinate, they produce one nucleus in each hyphal cell; the hyphae produced by the spore ultimately form the primary mycelium. This gives the fungus the same basic genetic build-up throughout and is termed <u>monocaryotic</u>. It is possible for the cells of some species to become binucleate by division of the nuclei, but even in such cases the mycelium retains the same genetic build-up throughout; such a mycelium is called <u>secondary</u>. If an agaric can complete its life-cycle without exchanging or gaining nuclei from another mycelium or mito- (asexual) or meiospore (basidiospore), it is called <u>homothallic</u>.

Generally, however, two spores or mycelia of different mating pattern, designated as 'A'&'B', or '+'&'-', must come together and transfer nuclei from one to the other to form secondary mycelia. The two nuclei divide vegetatively (mitotically) within this mycelium at a scheduled time such that both nuclei can travel into each new cell produced thereafter. The mycelium produced is dicaryotic. This phenomenon, where species can only produce fertile basidiomes if both components are present, is called <u>heterothallism</u>. It is a common feature of the agarics.

Some mycologists believe the clamp-connection was evolved in order to obtain the correct complement of nuclei in a cell (the clamp-connection is described on pg. 77 and illustrated in Plate 10, fig. J). This suggestion, if correct, is probably not the whole story, as many agarics have subsequently lost the ability to produce clamp-connections but nevertheless function apparently normally.

Unfortunately only a few species have been used to study the genetics of the agarics, and it is rather unfortunate that those that have been documented are not always typical of the group (eg. *Coprinus cinereus* and *Schizophyllum commune;* indeed the latter is probably not a true agaric at all). The study has been confused still further by misidentification of the subjects of experimentation, eg. *Coprinus cinereus* has often been called by the name *C. lagopus* which is quite a different fungus.

By analysis of the genetics of agarics such as *Coprinus cinereus* it can be demonstrated that in order to procure new genetic material outbreeding is favoured. In parallel to the results obtained by inflicting cultural taboos on humans, mating between related fungal individuals through the course of time is thought to be disadvantageous. When the two nuclei fuse in the basidium, as described above, chromosomes divide and genes exchange; this is followed by a division of the zygote to produce two nuclei. Each of these nuclei further divides but without exchange of genetic material; it is only now that the nuclei migrate into the four spores which form at the apex of each basidium. Generally only two of the four are alike in their mating pattern and cannot fuse; mating can only take place between one of each. An agaric by so doing attempts to outbreed and is called bi-polar.

An extension of this pattern of behaviour is found in more complex agaric-groups where each of the four spores is unable to mate with the other and therefore must go outwith the parent basidium to find a suitable partner. Such an agaric is called tetra-polar. The two factors involved in producing this tetra-polar pattern are called A & B factors; when only common B factors are present, no mating takes place. Finally, and only recently, an even more sophisticated outbreeding system has been demonstrated in members of the genus *Psathyrella*; this is termed octa-polar. Only now do we possess sufficient genetic information to carry out studies on edible mushrooms parallel to those of plant-breeders and so improve stocks etc.

The typical text-book agaric is tetra-polar and possesses clamp-connections on the vegetative hyphae. To those familiar with genetics, such phenomena are characteristic therefore of a fungus which has two mating types on different chromosomes, with both loci usually possessing multiple alleles.

Bi-polar species are probably as common as tetra-polar species in nature, and confusingly they may or may not exhibit clamp-connections on their vegetative hyphae. To use the jargon of geneticists, they have a single mating-type locus with multiple alleles.

Monocaryotic hyphae which develop from spores borne on a four-spored basidium lack clamp-connections, but are frequently festooned with asexual propagules called meiospores, or conidia; these conidia are described on pgs. 66 & 70, and are in general of the thallic arthric type. Sometimes spores produce hyphae which immediately exhibit clamp-connections in the vegetative stage. Further examination shows that these spores were formed on two-spored basidia; that is half the number of spores the basidium of a 'normal' agaric possesses. Every agaric has a few such basidia, which often do not produce viable spores, but some species are characteristically two-spored. Such fungi are called <u>secondary homothallic</u>; that is, each spore contains one nucleus of each of the mating types (eg. *Mycena galericulata*, *Conocybe blattaria* & *Agaricus hortensis*).

Some agarics with two-spored basidia do not produce clamp-connections, and it is only by experimental analysis that the genetic build-up can be understood. It is hypothesised, both on morphological and cultural characters, that the 2-spored species have been derived from 4-spored species.

Dicaryotic hyphae usually exhibit clamp-connections. Some species are constantly without them, or they form clamps in particular parts of the lifecycle (eg. base of the basidium in *Armillaria bulbosa*, or hyphae at the base of the stipe, eg. in *Chroogomphus (Gomphidius) viscidus*).

If an agaric can be cultured, the following simple exercise can be carried out to ascertain its characters. Monocaryotic cultures can be prepared by picking off single spores (see pg. 56) and producing a colony. Dicaryotic hyphae can be obtained from basidiome tissue. If the spores fail to germinate but a tissue culture (see pg.31 & 32) is available, then maceration or chemical dicaryotization can be conducted (see pg. 57). By careful maceration even the two different nuclei in the cells of hyphae produced by spores from two-spored basidia can be separated. This allows the analysis of previously difficult species to be accomplished.

2. MATING OF KNOWN AND UNKNOWN STRAINS

Place two monocaryotic inocula about 10 mm apart on a Petri dish so that nuclear migration can be studied. The two should be checked when they commence their new growth to ascertain that they are indeed pure monocaryons; cultures sometimes dicaryotise because spores become enveloped in the growing mycelium and fuse with it, then later germinate. When the two mycelia placed 10 mm apart come into contact, clamp-connections are rapidly formed throughout the colonies if the strains belong to the same species. If the two inocula are truly monocaryotic, and unilateral nuclear migration is recorded by the formation of clamp-connections in one of the cultures only, then these are also to be considered belonging to the same species. The test needs to be repeated several times to overcome the effect of heterothallism see pg. 92. Migration is when the nuclei move along the hyphae of the inoculum to make the paired complement. Two-spored species in nature lack clamp-connections, and so detailed studies between strains of the same species from many widely scattered areas must be undertaken to decide whether the taxon is heterothallic or not. Maceration techniques are utilized: see pg. 57.

If development of clamp-connections does not occur within ten hours of meeting, then inocula should be taken from the junction line between the two mycelia and grown on new agar surfaces. In the culture so produced, the formation of ill-formed, non-fusing, backwardly oriented branches, called false clamp-connections, (Plate 10, fig. V) and a rather streaky mycelium suggests that the Common -B factor of a tetra-polar species is present. Further pairings of other strains should produce some cultures which have no mating types in common, and therefore good dicaryons will be formed.

Common -A and Common -B factors are found in tetra-polar species, as indicated on pg. 92; mating of colonies with A & B factors is not operative in bi-polar species. In this case, the mating of one strain of an unknown with one or other of the strains of the known must give a single compatible mating if the isolates are to be considered to belong to the same species.

If both strains happen to be 'blockers', ie. a distinct reaction against one another is found, the dikaryotic mycelium can be often only isolated from the junction line. "Blockers' are factors which halt the movement of nuclei through the hyphae. Care must be taken in analysis of these agarics, which should be attempted only after experience has been gained and more advanced texts consulted.

Negative results on their own are useless, and ideally it is necessary to work out the incompatibility system of the known and unknown isolates before deciding as to the non-crossability of two intra-fertile groups of strains. At least now the agaricologist has available a powerful tool in his attempts to identify unknown cultures.

3. MATING-TYPE DETERMINATION

Prepare single-spore cultures of the agaric under study and allow them to grow for 2-3 days. From the actively growing margin of these cultures, cut out plugs with a flamed cork-borer 1-2 mm^2, and place them in pairs on a clean plate 10 mm apart. Leave in a clean place at room temperature for 24 hours before examining them for the presence of clamp-connections; score the results from a dozen confrontations. In this way one can determine whether an agaric is tetra-polar or bi-polar. If the formation of clamp-connections occurs in approximately one quarter of the cultures the fungus is tetra-polar and if approximately in one half of the cultures it is bi-polar. In tetra-polar species false clamps (see pg. 93) are often formed; always check therefore that fusion takes place in the arm of the clamp-connection.

It is now illustrative to repeat the experiment using basidiomes of the same or other species. Single spore colonies can be obtained either from random spore platings or directly from tetrads (see pg. 35).

The following type of tables will be obtained:

Tetra-polar									Bi-polar								
Isolate number	1	3	4	6	7	8	2	6	Isolate number	1	2	4	7	3	5	6	8
1	–	–	–	–	–	–	+	+	1	–	–	–	–	+	+	+	+
3	–	–	–	–	–	–	+	+	2	–	–	–	–	+	+	+	+
4	–	–	–	–	–	–	+	+	4	–	–	–	–	+	+	+	+
5	–	–	–	–	–	–	+	+	7	–	–	–	–	+	+	+	+
7	–	–	–	–	–	–	+	+	3	+	+	+	+	–	–	–	–
8	–	–	–	–	–	–	–	–	5	+	+	+	+	–	–	–	–

```
2      + + + + + − − −           6      + + + + − − − −
6      + + + + − − −             8      + + + + − − − −
```

Successful matings can also be observed by changes in growth rate of the colony, floccosity of the culture, production of pigment, etc., as outlined earlier (pg. 75) under the heading 'Additional culture characters'.

4. HOMING RESPONSE: ITS USES IN CULTURAL STUDIES

Many species of agarics produce conidia (see pgs. 66-70); in the genus *Coprinus* about 75% of the species produce such propagules. In common with many agarics, the propagules found in species of *Coprinus* are thallic arthroconidia, as outlined on pgs. 62-63.

Conidia seem to be of two basic types, wet and dry. 'Wet' conidia coalesce into a droplet, and dry conidia fall onto the surface of the agar as a powder after break-up of the chain of cells. Both 'wet' and 'dry' conidia can germinate, but the 'wet' type is often poor at germinating, whereas dry conidia do so immediately under favourable conditions.

The thallic arthroconidia of many species function sexually not only by simply landing on monocaryotic hyphae and fusing with them but also by producing substances which can be recognised by the hyphal tip so that growth occurs in the direction of the conidia. These conidia are termed <u>spermatia</u>. This atrraction has been called the <u>homing effect</u> or <u>homing response</u>. It can be demonstrated by spreading a suspension of arthroconidia on half of the agar surface in a Petri dish and a small plug of inoculum of the same species placed in the other half. All the hyphae become more or less oriented in the direction of the conidia. Fusion soon follows, and with monocaryotic cultures, clamp-connections are formed if the conidia and hyphae are of opposite mating types. Both 'wet' and 'dry' conidia can cause homing, but 'wet' types are poor at germinating and good at causing homing; in contrast, dry conidia have often the reverse properties.

The response to the presence of conidia is not species-specific and can occur between a hypha and a conidium of closely related species; indeed this test can demonstrate the closeness of two species which may not be at first obvious from morphology of the basidiome alone. When homing occurs between different closely related species, fusion is usually followed by death of the cells in the hypha; this can be seen by the hyphal tip becoming bubbly inside. It is a lethal fusion.

Between distantly related species, fusion does not take place. The dispersal of conidia by existing monocaryotic hyphae can therefore suppress the growth and establishment of a mycelium of a late-comer to the habitat. Tests of homing and lethality between hyphae and thallic arthroconidia acting as spermatia can eventually be used to confirm the identity of a species of agaric. It is a very powerful both genetic and taxonomic tool.

The fusion of conidia and hyphae often takes as little as 1 hour for completion and therefore can be followed under the microscope. By picking off a single arthroconidium in the same way as described for single spores (pg. 56), a conidium can be transferred to a position close to a hypha. Fusion will take place, but some agarics fuse using a peg from the leading cell, others dichotomise and capture the conidium, whereas others turn around and form a crook by which the conidium is surrounded (see Plate 12: lower figs.).

5. COMPARATIVE STUDIES

All the experiments above can be repeated for cultures of *Panellus stipticus*. Bioluminescence is probably governed by a single gene in this species, or by a group of closely interconnected genes. The pattern of bioluminescence is found only in N American strains, and the character can be followed in different experiments including the formation of the hybrid strain.

Within the genus *Coprinus*, the sections appear to be fairly homogeneous in the mating-pattern behaviour of their constituents. Thus all common members of the sect. *Lanatuli*, ie. *Coprinus cinereus* group, are tetra-polar, produce wet arthroconidia, and exhibit clamp-connections on their hyphae. (See Kemp, Appendix IV)

Members of the sect. *Setulosi*, however, are more variable and demonstrate the whole range of patterns available. Different species within the section should be cultured to appreciate this variability. Some species are homothallic, others bi-polar, and yet others tetra-polar. The presence of clamp-connections differs with the species as does the production of conidia, and when present the conidia formed may be dry or wet depending on the species. Some species, eg. *C. sassii* and *C. bisporus* are characterised by 2-spored basidia.

6. PHYSIOLOGICAL EXPERIMENTS

Much has been written on the physiology of the larger fungi in culture, and by now an amazing amount of information has been amassed on a few particular species. Such studies are out of the scope of this text. Many experiments have been carried out on the vegetative state of agarics, but there is also a significant number of studies on the effects of light, growth substances, atmospheric content, humidity, etc. on the production of basidiomes.

Several media have been designed especially for studying the various aspects of growth and sophisticated techniques developed to monitor changes in growth, eg. media which can be irrigated with nutrient solutions, or in which the nutrients can be changed or diluted.

Development has been studied particularly by the use of genetic markers so that even small changes can be monitored. For such techniques the reader must consult more advanced texts. A few important references are given in the reading list in Appendix IV.

VI. SUGGESTED EXERCISES

A. LABORATORY EXPERIMENTATION FIG. 12

Singer (1961) has reviewed many of the books which cover the techniques of isolation and growth of the mycelium and the production of fruit-bodies of those species of <u>Agaricus</u>, <u>Volvariella</u> and <u>Lentinula</u> grown for human consumption. His work supersedes earlier commercial and government bulletins which frequently only deal with <u>Agaricus hortensis</u> s. lato, eg. Duggar (1905); Atkins (1956); Jackson (1951): see refs. in Appendix IV.

Treschow (1944) has considered certain physiological aspects of mushroom culture adequately, as has Mader (1943). Further information may be sought in articles appearing in <u>Mushroom Growers' Journal</u>, <u>Mushroom Science</u>, and other similar publications. Recent issues cover not only *Agaricus* but also *Pleurotus ostreatus*, *Flammulina velutipes*, and *Agrocybe cylindraca*. A recent compilation which covers the production of *Flammulina velutipes*, and *Pholiota nameko* in addition to the more familiar edible species is edited by Chang, S. T. and Hayes, W.A. and is entitled <u>The biology and cultivation of edible mushrooms</u>.

Several of the edible agarics prove to be useful not only to study development of the basidiome but as experimental material, mainly because the species grow relatively easily in the laboratory.

1. *Agaricus hortensis* and *A. brunnescens*
2. *Coprinus cinereus*
3. *Stropharia ferrii*
4. *Psilocybe cubensis*
5. *Pholiota nameko*
6. *Galerina mutabilis*
7. *Agrocybe cylindraca*
8. *Volvariella esculenta*
9. *Flammulina velutipes*
10. *Pleurotus ostreatus*
11. *Lentinula edodes*
12. *Schizophyllum commune*

1. *Agaricus hortensis* (Cooke) Pilat and *A. brunnescens* Peck (see Plate 13)

These agarics are known under several epithets. However, it is best to retain the names used herein for the white and brown forms respectively. If they are found to be the same species, then the latter has precedence. They are probably all derived from *Agaricus bisporus* (J. Lange) Pilat, the wild species found in hedge-row bottoms in Europe. Apart from a few records of escapes *A. hortensis* is known only in cultivation.

'Home-Grow kits' are now commercially available at a moderate price and reduce the necessity to isolate and produce spawn in the laboratory. The results from these kits are more reliable than trying to procure one's own basidiomes by pure culture methods, unless good laboratory facilities are available. Much research work has gone into attempting to fruit *Agaricus* in the laboratory in pure culture, but so far consistent attempts have failed. The directions on the label of the kit should be followed, but normally *Agaricus hortensis* can be grown at room temperature in the light or dark. About four weeks are necessary before primordia will be seen, although growth can be inhibited by the accumulation of carbon dioxide in the culture. After maturation of the primordia new crops will develop approximately weekly.

Tests

After cutting off the stipe of a mature basidiome and fixing the pileus to the top of a tall glass vessel, the spore drop can be observed by viewing from the side whilst a strong beam of light is shone through the vessel. Spore-prints collected at the bottom of the vessel will show paired grouping: see pg. 57. Spots of Indian ink placed equidistant along the stipe and pileus when the primordia are 50 mm high allow the expansion of the basidiome to be followed by measuring the change in distance between the spots. This shows that the pileus expands at the margin, and the stipe elongates at both the base and the apex but more particularly above the ring (Plate 13, fig. A).

Make parallel cuts in a young basidiome such that the pileus extends from only two sides of the basidiome to make a 'T' shape, and then remove the gills from one side. As further growth takes place, the stipe will bend away from the side bearing the remaining gills, indicating the activity of hormone-like compounds.

Similar curvature can be induced by placing agar blocks, containing either acetone or ethanol extracts of the gills, unilaterally against a basidiome freed of gills. Amino-acids in dilute solution have similar results. See Plate 13, figs. B-L. Additional experiments which are instructive involve the insertion of a piece of inert material, eg. mica flake, into a cut in a developing basidiome. The subsequent curvature depends on where and at what time the cut was made: See Plate 13, figs. M & N.

Increase in size of the basidiome involves not only up-take of water but an increase in dry matter. This can be demonstrated by measuring the wet and dry weights of basidiomes at different stages in development and plotting the results graphically. A more or less linear plot if obtained. Applying the techniques described in the earlier text, cultures and basidiomes can be obtained in the laboratory. Some commercial strains of *A. hortensis* produce primordia on malt agar, but the number of primordia between strains and within each strain is modified by the source of the agar, pH of the agar, presence of additional carbohydrate sources, and accumulation of carbon-dioxide.

Fruiting can be encouraged by covering the culture with a layer of activated charcoal, which apparently removes self-inhibitors of fruiting.

Always try and use the cultures prepared on agar surfaces as soon as possible, as continual transfer to fresh agar medium encourages the formation of 'spots' with different genetic constitution (saltation). Transfer the culture to a glass vessel containing about 100 gm of sterilised rye grain; barley, rice, wheat or corn grain obtained from a food store will be equally suitable. A pre-treatment of a 12 hour soaking of the grain in 150 ml of distilled water is preferable. Any container can be used, provided similar proportions of grain to water are used, ie. 2 parts of grain to 3 parts of water in approximately equal parts by volume. Sterilise the containers for 45 minutes with their tops slightly loose, and then allow to cool in the steriliser or pressure cooker; tighten the lids until required for inoculation.

The inoculum can be prepared from spores or tissue as outlined on pgs. 31-35. It should be introduced into the system as small pieces as opposed to a single large block. Try and disperse the inocula through the grain by shaking. Mycelium grows rapidly over the grain, and within twenty days at room temperature it completely covers much of the grain particles. The grain itself when covered in mycelium can then be used for seeding compost or additional glass jars containing grain.

A fully colonised grain sample should then be covered in a casing soil to stimulate growth of mycelium and production of primordia. The soil should be a mixture of peat (40 g), ground chalk (1 g), vermiculite or loam (20 g), and fine sand (20 g). The last three constituents after thorough mixing should be sterilised at 15 lbs pressure for 45 minutes. Peat sterilised at this temperature tends to liberate noxious materials and so should be sterilised at 10 lbs pressure for 10 minutes, a procedure which must be repeated five times. The peat should be then washed in a sterilised sieve with sterile water before mixing with the other constituents. Enough casing soil should be added to the grain to cover it to a depth of 3-4 cm. Once inoculated the culture should be left on a window-sill at room temperature. Within two to three weeks the mycelium will 'rin', ie. colonise the casing soil, and form primordia. The casing soil should be damp when added to the bottles; this level of moisture should be maintained throughout production of mycelium and basidiomes by carefully removing the lid and spraying onto the surface sterile distilled water from a syringe.

The use of compost prepared by replacing the grain by cut pieces of straw see pg. 8, has been successful. Techniques such as these can be used for the production of basidiomes of *Stropharia ferrii* (see below) and *Psilocybe cubensis* (see below), giving usually much better results than with *A. bisporus*, which tends to be erratic in the production of its basidiomes. *Agaricus bitorquis* (Quelet) Sacc. can be substituted with success for *A.*

hortensis and would appear to give much more consistent results. Isolates prepared from other wild species should be tested in a similar way.

2. *Coprinus cinereus* (Schaeff. ex Fr.) S.F. Gray

Coprinus cinereus occurs on straw-bales, dung/straw mixes, and piles of farm-refuse. It is very common, although frequently misidentified and confused with large basidiomes of *C. radiatus* (Bolt. ex Fr.) S.F. Gray. It has been called *C. lagopus* (Fr.) Fr. by many geneticists and appears as such in several journals and books. *C. lagopus* differs in its preference for woodland habitats and in several microscopic characters. The *C. lagopus* which appears in Buller's classic publications on the biology of larger fungi is in fact *C. radiatus*.

C. cinereus is quite variable and in addition has been isolated from substrates as wide apart as fermenting grass and human heart-tissue; it is thermotolerant. Those basidiomes with a long rooting base have been called *C. macrorhizus* (pers. ex Fr.) Rea. The name *C. fimetarius* (L.) Fr. has been applied to both *C. cinereus* and *C. radiatus*.

Prepare cultures either from a multi-spored source or from a previously mated compatible strain. These should be then inoculated into sterilised horse dung kept in an open tray or in a crystallizing dish. The dung is best incubated in the light at $25°C$; simply placing on the window-sill is usually adequate. Although the primordia do not require light for initiation, the basidiomes require light for further development.

Primordia are produced within ten days of inoculation and basidiomes in two to three weeks. As older basidiomes deliquesce, more basidiomes will develop, giving a reasonably reliable source of basidiomes in all stages of development.

C. cinereus may also be fructified on such defined media as described on pg. 6, and so cultivation can be achieved in Petri dishes without resorting to the use of dung.

Removal of the fruiting cultures to a refrigerator will prevent autolysis for a while, perhaps even for as much as 8 hours, but basidiomes stored in the cold for more than an hour or so rarely produce mature basidiospores. From a single culture not only the development of the basidiomes can be followed but also the process of autodigestion and spore-release studied. The elongation of the stipe of *Coprinus cinereus* is a very rapid process, with stipes elongating from 10 to 100 mm in a few hours. The results of experimentation over widely differing nutritional states suggest that chitin synthesis is an essential process in this elongation.

C. cinereus and many other agarics can be induced, although erratically, to produce monocaryotic fruitings by growing the monocaryons on agar in a Petri dish and irrigating the culture with a cell-extract from a dicaryon of the same species. The cell-extract can be obtained by macerating the culture and milipore-filtering the resultant suspension. Monocaryotic cultures of *C. cinereus* grown on a permeable membrane seated directly on the dicaryon will give similar results.

Both simple and sophisticated genetic studies can be carried out using *C. cinereus*: see pg. 90. Basidiospores from freshly autolysed basidiomes or fragments of gill for later comparative studies can be stored in screw-cap vials filled to a depth of 1-2 cm with sterilised, coarse silica gel. Sterilization of the silica is best achieved using a hot air oven; 100-mesh gel is preferable. Thallic arthroconidia may be stored as a suspension in water in a concentration of about 1×10^6 ml and mixed with an equivalent volume of sterile non-fat milk. Alternatively, approximately 0.25 ml of such a suspension can be added to the silica-gel as above and stored at 4^oC.

Cultures can be stored at 4^oC in screw-cap phials with moisture-tight liners, or cotton-plugged and covered in 'Parafilm'. For longer-term storage see pg. 44.

C. congregatus and many other members of subgenus *Pseudocoprinus* section Setulosi grow well in artificial culture. Basidiomes produced in continuous light are characterised by not producing basidiospores and not autolysing, and by having reduced stipes. The formation of a normal basidiome requires light and dark periods, and once formed, the presence of a pileus is necessary for complete stipe-elongation. Growing specimens in the dark at 25^oC followed by a cold, light treatment (10^oC) gives normal basidiomes, but the reverse is not true. The stipe-elongation and development of the pileus can be achieved in light-initiated primordia by growing the culture at 10^oC in the light. Primordia develop at the expense of glucose, and when this is depleted further growth and elongation ceases.

3. *Stropharia ferrii* Bres.

First described in 1926 from Europe as *Stropharia ferrii* Bres., this very handsome agaric was forgotten, even though in the United States Farlow had described it again, as *S. rugosa-annulata* three years after Bresadola. It has probably been introduced to Europe from North America, where it is fairly common and widespread, especially on mulch around amenity trees. It has been placed in both *Naematoloma* and *Geophila*.

This fungus grows very well on a sawdust medium, and if the procedure used widely for the cultivation of polypores is followed, the formation of primordia will be induced. Badcock's medium, the recipe for which is outlined on pg. 10, is probably the most useful medium; a procedure similar to that used for *Pleurotus ostreatus* is adequate (see below). Development of primordia takes about 7 days, and examination of the young developing basidiome allows the formation of chrysocystidia to be followed and the development of the velar tissue to be examined and plotted.

The spawn can be obtained commercially in Germany and the Low Countries, although it has become neither popular nor widely grown. Germination of basidiospores is high and easily carried out by using the techniques described in the earlier part of the book. A suitable inoculum can therefore be easily obtained and utilised as described on pg. 100 but by replacing the grain with sawdust.

4. *Psilocybe cubensis* (Earle) Singer

This species is placed in the same family as *Stropharia ferrii*. It was once placed in *Stropharia* because of the annular veil and may be found in this genus in many popular books even today. It differs from species of *Stropharia* in the absence of chrysocystidia in the hymenial tissue, ie. sterile cells which contain an amorphous yellow inclusion when mounted in aqueous ammoniacal solutions. *Psilocybe cubensis* and several related species grow well on a whole series of agar media. The most useful is malt agar, either with or without the addition of yeast extract (5 g per 1000 ml), or with or without mineral salts (see Kauffman's agar pg. 4). Inocula can be obtained either from basidiospores, as they germinate easily, usually within 3-4 days, or from tissue, see pgs. 31-35. The spores grow well in Petri dishes where small primordia form; better results are obtained by inoculating either paper-pulp or corn-meal media, to which may be added casing soil; see pg. 100. Some authors have used rye-grain supplemented with yeast extract (2 g per 100 g of grain), calcium carbonate (2 g), dipotassium phosphate (0.2 g), and .01 ml of a trace-element solution.

Some books on hallucinogenic fungi are accompanied by samples of basidiospores of *P. cubensis*, but these should be examined first to ascertain whether they are truly *Psilocybe* spores or not.

5. *Pholiota nameko* (T. Ito) S. Ito & Imai: 6. *Galerina mutabilis* (Schaeff. ex Fries) P. D. Orton

P. nameko was first described as *Collybia nameko* and has been placed previously in *Kuehneromyces*; see below. Singer places this species in a separate section of *Pholiota* sg.*Hemipholiota* based on its glutinous veil.

The fungus is found in Japan on dead trunks and stumps of deciduous trees, especially of beech and oak. Basidiomes can be produced quite easily in the light on a sawdust mixture prepared in a similar way to Badcock's medium; see pg. 10.

Galerina mutabilis, also sometimes placed in *Kuehneromyces*, is probably more commonly known as *Pholiota mutabilis*; it will also fruit using a sawdust mixture or a paper-pulp medium.

Both fungi will allow the bivelangiocarpic development to be followed, glutinous in the first species and dry and fluffy in the second. In *Pholiota nameko* the pileus is smooth and covered with glutinous mucilage; it is not hygrophanous. *Galerina mutabilis*, also commonly given the popular Japanese name 'nameko', has a humid, hygrophanous pileus which commonly appears belted and quite yellow-ochraceous in the drying areas. In this same species the stipe is also roughened, with tawny to vandyke-brown scales terminating $\frac{2}{3}$--up from the base in a little feathery-scaly ring. The ring is thin and fugacious in *P. nameko* also, although at first it is membranous and glutinous. The stipe is roughened-floccose below the ring.

Both species are caespitose, and observations in culture demonstrate the tight cluster of basidiomes interlocked to form a tuft. In other species of agaric, all but a single primordium in a cluster are inhibited, and so only solitary specimens develop.

7. *Agrocybe cylindraca* (DC. ex Fr.) Maire

This species was once placed in *Pholiota* because of the presence of an annulus (ring) and brown spore-print; it is more widely known as *Agrocybe aegerita*, although it can be found in the literature under the synonyms *Pholiota cylindracea* (DC. ex Fr.) Gillet and *Pholiota pudica* (Bull. ex Merat) Gillet.

This species can be isolated from spores or tissue and fructifies in the same way as *S. ferrii*. It has a typical cellular pileipellis and bolbitiaceous hymenophoral trama and spores. Although more time-consuming, successful culture has been achieved by adopting the techniques used in Japan for Shiitake; see below.

8. *Volvariella esculenta* (Mass.) Singer

This is a member of the Pluteaceae; it is also known under the name *Volvaria*. It is cultivated in southeast Asia as an edible fungus and is probably best known as the commercial paddy straw fungus.

Take wheat or rice-straw or similar grass-stems and chop them into small lengths. Soak with water and leave in open boxes until natural break-down occurs. Break up some old basidiomes of the mushroom and sprinkle on the straw as the upper layer is turned, or use pure cultures and inoculate below the surface layer. Within 3 weeks basidiomes will form, and although some may develop and mature, the rest can be encouraged to go to completion by covering the straw with casing soil 3-4 cm deep; see pg. 100.

Volvariella is distinguished by the presence of a volva. A large number of primordia are formed, and therefore the formation of the pileus within the volvate tissue can be followed. The development of the inverse hymenophoral trama can be reconstructed from the examination of serial sections, processed as described on pages 52-54.

9. *Flammulina velutipes* (Curt. ex Fr.) Karsten

This species grows in clusters on trunks and stumps, frequently on weakened but living trees. It fruits chiefly late in the season, appearing from cracks in the bark or from decaying stumps; it causes a white rot of the sapwood. This agaric is probably better known under the name *Collybia velutipes* (Curt. ex Fr.) Kummer. It has been placed in both *Myxocollybia* and *Pleurotus*.

Take a fresh basidiome and prepare a spore-print as outlined on pg. 35. Shake up the spores in water and inoculate a Petri dish containing agar medium. Incubate the medium at $25^{\circ}C$ for one week in the dark, and then move to a cool illuminated incubator at $15^{\circ}C$. Mature basidiomes appear four to five weeks after inoculation of the basidiospores. As in *Coprinus*, light is not necessary for initiation of the primordia, but it is a necessary prerequisite for the expansion of the pileus. Initiation of the primordia and their subsequent growth are inhibited by the build-up of carbon dioxide in the culture vessel. Cap-less basidiomes can be obtained by maturing the primordia in the dark, but no stipe-elongation ensues even if the culture is later brought out into the light. Good fruiting can be obtained by incorporating poplar sawdust or a similar wood base into the medium. *Flammulina velutipes* has been used for studies on growth patterns, as very small numbers of lamellae will produce detectable amounts of an active diffusate; the diffusate can be increased by supplying nutrients to the excised lamellae. The activity is greatest close to the onset of rapid elongation. The pileus influences the growth of the stipe, probably by acting as a source of growth substance.

Whole detached basidiomes respond if in the young state but less significantly towards the end of their rapid growth stage. During the period that the stipe is dependent on the pileus, three quarters or more of the normal stipe growth can be restored to detached whole basidiomes by feeding with glucose or with various natural extracts. The function of the pileus can only be partially replaced by diffusates of lamellae (=gills) and never by nutrients. In young pilei the development of the pileal cystidia can be successfully followed in culture.

10. *Pleurotus ostreatus* (Jacq. ex Fr.) Kummer

This fungus is widespread on standing dead and dying timber of both frondose and coniferous species. It is characterised by a pinkish to lilaceous spore-print and has been consistently confused, particularly in North America, with *P. cornucopiae* (Paulet ex Pers.) Rolland (= *P. sapidus* Schulzer apud Kalchbr.) Sacc. which has a white spore-print. *P. ostreatus* has also been known as *P. salignus* (Pers. ex Fr.) Kummer, and a blue-coloured form known as *P. ostreatus* var. *columbinus* Quélet is not uncommon on poplars in Europe. In North America there would appear to be other closely related species.

Basidiospores obtained from spore-prints of any one of these taxa germinate readily, and if a multi-spored inoculum is placed on paper-pulp or sawdust medium, primordia develop within 3 weeks. Maturation of these primordia develop within 3 weeks. Maturation of these primordia is a little more erratic, but within 7 days after initiation of the primordia, maturing basidiomes are usually formed. At first the stipe is central, but gradually there is unilateral development of the pileus to produce the more familiar oyster-shell shape.

Similar results are obtained with the Asiatic *P. flabellatus* and *P. fossulatus*.

Pleurotus species grow and fructify consistently on a mixture of cornmeal (24 g), corn starch (24 g), wheat bran (24 g), oak sawdust (24 g), and malt extract (8 g) added to 180 ml of water dispersed in glass vessels and autoclaved for 20 minutes at $121^{\circ}C$. Mycelium grows rapdily, but fruiting is erratic unless a dark/light regime is introduced. The following growth-chamber pattern has been used at Blacksburg, Virginia, N. America: $20^{\circ}C$ (12 hr) day/$15^{\circ}C$ (12 hr) night at about 60% relative humidity and fluorescent lighting: see Appendix IV.

It is usually necessary to keep the growing primordia bathed in aqueous solutions. Covering the initiated primordia with peat moss and daily watering assists in allowing the fully mature basidiomes to develop. Soil, activated charcoal, etc. can be substituted.

Consistent fruiting of *Pleurotus ostreatus* has been obtained by growing cultures, taken from either tissue or basidiospores, in the dark, and after their establishment transferring them to malt-agar under red-light conditions. Growth should be allowed to continue for 14 days, during which time cultures are irrigated with a mixture of 1 ml of 0.05 M sodium phosphate buffer and

2% solution of asparagine, and illuminated with white light. The cultures are then maintained in the light at 20-25°C.

Attempts are now being made to determine whether *P. ostreatus* is a single variable species or several closely related autonomous species. Bresinsky and colleagues (1977) have paid particular attention to European collections, whilst Egger (1978) and Eugenio & Anderson (1968) have surveyed several North American strains. The picture, however, is still far from clear; see refs. in Appendix IV.

Parallel work is now underway on members of the *Armillaria mellea* complex. Undoubtedly this name also covers a group of closely related taxa, but it is difficult by conventional means to separate them clearly. The work of Korhonen (1977) in Finland and Ulbrich and Anderson (1978 & 1979) in the United States shows how such problems can be tackled using the techniques outlined in this text.

11. *Lentinula edodes* (Berk.) Pegler

This is another of the pleurotoid group; it is undoubtedly better known as *Lentinus edodes*, or 'Shiitake'; in some texts it has also been called erroneously *Cortinellus edodes*.

Small kits of this species can be obtained on the European market as the 'Black Forest mushroom'. Even though one carries out the instructions on the label, fruiting is erratic and not as predictable as indicated by the manufacturers.

Normally, in Japan where it is grown commercially, small plugs of the vegetative state are hammered into pre-formed holes drilled in oak branches previously cut into 1 metre lengths. These lengths are then stacked in shady places and watered, and 3-4 months later the basidiomes begin to appear. Depending on the success of the spraying, basidiomes can be expected to appear for several months.

In this species a secondary veil is present, the development of which can be followed by examining the basidiomes in various stages of formation.

12. *Schizophyllum commune* Fr.

Schizophyllum is not an agaric in the strictest sense but is more closely related to members of the Aphyllophorales. This agaricoid species grows on vegetable material, fallen twigs, branches, logs, stumps, etc.; it is world-wide. It has also been isolated from human sputum, cerebral fluid, mouth ulcers, and a whole range of manufactured goods from vegetable dyes to building materials.

Schizophyllum commune is unusual in development, and many correlated characteristics make it unique, especially the feature of it retaining the ability to produce basidiospores even when the basidiome has been dried for very long periods. A source of spores therefore can be stored in the form of air-dried basidiomes in paper envelopes or boxes. When cultures are required, moistened basidiomes are set over agar surfaces as outlined on pg. 36 and the spores allowed to fall onto the medium.

Basidiospores obtained in this way will germinate within 24 hours, and small colonies should be expected within 3 to 4 days. If dikaryons have been produced, then fruiting will take place after a further 7 days. Fruiting takes place in Petri dishes or in any other similar vessel if the culture is incubated at 22°C in the light. Fruiting appears to be enhanced by making incisions in the agar.

Stock cultures can be maintained for about 2 years if treated as outlined on page 44. Light mineral-oil preservations is useful.

The polymorphism of *S. commune* can be demonstrated by standing some cultures on their side and others upside down. The basidiomes of *Schizophyllum* commence as a globose primordium with an apical cavity in which the hymenium develops; further development continues as marginal in-growths of the pileus. In those cultures on their side, the pileus expands unilaterally and the stipe appears excentric. In other cultures, the stipe is attached to or near to the centre.

Schizophyllum has been used for many sophisticated genetic studies, and the reader should refer to these. They are outwith the scope of this text.

B. Ecological exercise - Coprophilous fungi

Coprophilous fungi are exceedingly satisfactory for demonstrating the diversity and morphological variation of a single group of related organisms within an ecological system. Representative genera of all the major groups of fungi can be guaranteed to appear on dung after a period of incubation. They have the added advantage in that coprophilous fungi are apparently cosmopolitan. Although in many countries the horse is fast disappearing as a working animal, there is certainly no shortage of dung in our fields and woods, and this material will always produce characteristic fungi whatever time of year it is collected. Dung of sheep and goat, hare and rabbit also make very suitable substrates for laboratory study; cow dung is less manipulable.

Dung is best incubated in a light place, for example on a window-sill in a warm room, on layers of filter paper or other absorbent material. For rabbit-pellets and samples of similar size Petri dishes are ideal, but for horse 'apples' and larger types of dung, large covered dishes such as glass casseroles or plastic sandwich containers are needed. Samples should not be kept in airtight containers for any length of time after collecting, as in such conditions insects and nematodes tend to break down the dung, and anaerobic conditions rapidly develop, which apparently encourage the disorganization of basidiospores. If the samples cannot be set to incubate soon after collection, they should be gently air-dried, for most dung fungi will remain alive after such treatment and grow out when the sample is eventually moistened. The filter paper on which the sample is placed should be kept moist, and although free water will not allow the best development of ascomycetes the succession of basidiomycetes appears to vary with the wetness of the dung. Samples kept in constant mist, with or without underneath heating, give interesting results deviating from the general pattern usually observed. Earth-worms and fly-larvae should be excluded from the samples as far as possible, for they break the dung up too much. Activity of the latter can be considerably reduced by spraying the sample very lightly with a proprietary fly-kill aerosol. If space is limited and pure cultures are kept nearby, it is very important to prevent transference of mites from the dung to the cultures. Under these circumstances the containers should be placed on glass plates lightly smeared with vaseline, to which should be added an acaricide, eg. methyl benzoate.

Agarics are best sought with a stereoscopic binocular microscope, when their full beauty will be seen, but a hand lens or simple magnifier, if less convenient, is sufficient. The larger ascomycetes and most of the basidiomycetes are readily seen with the naked eye, but the binocular microscope is still very useful for observing the gross features of the veil in many of the latter group. Basidiomes of many of the coprophilous agarics, because of their size and/or rapidly deliquescent nature, often do not give good spore-prints. Mature spores in such cases should be sought on the stipe or in natural spore-prints formed on the filter paper. These spore-prints can be used for further culture studies.

A key to the more common basidiomycetes found on dung and associated debris is provided by Richardson and Watling (see refs. in Appendix IV) and can in addition be used for fungi growing in mushroom beds. An introductory part of the key indicates some of the other groups of basidiomycetes apart from agarics which may be met with. As the agarics have always been thought of as the best known of the coprophilous fungi, attention to their taxonomy and nomenclature has often been careless. In their key, Richardson and Watling have utilised a rather narrow species concept and in certain places indicate where further distinct taxa might be expected.

On any one dung sample a representative or even several representatives of all the major groups of Coprini may usually be found. The variability of the veil in the genus Coprinus can be easily demonstrated by picking off the veil, if one is present, from maturing basidiomes and examining them under the low power of a microscope; some species lack a veil. The velar remmamts should be mounted directly in water or ammoniacal solutions for microscopic examination. When these same velar remnants are placed on the surfaces of cultive media, vegetative colonies are often produced; velar tissue is apparently admirable for such work. If a veil is not present, then the entire developing basidiome may be placed under the low power of a microscope and the outer surface of the pileus examined. In this way the presence or absence of dermatocystidia may be demonstrated. Plate 11, figs. F-L illustrates those structures which can be observed by these simple techniques.

Not only is dung culture a useful way of obtaining basidiomes in fresh condition for later study, but studying fungi in dung-culture is a fascinating pastime.

Examples: (Plate 11)

a. Naked pileus-surface—Coprinus miser and allies, ie. Sect. Nudi. Fig. L.

b. Pileus-surface with cystidia—Coprinus bisporus, C. congregatus, C. pellucidus, C. sassii, C. stellatus, ie. Setulosi. Fig. K.

c. Veil of sausage-shaped elements—C. cinereus & C. radiatus, ie. Lanatuli, Fig. F.

d. Veil of round elements—C. ephemeroides, C. cordisporus and allies (acid-soluble ornamentation on cells. Fig. H): C. narcoticus, C. stercoreus & C. velox (acid-fast ornamentation on cells: Fig. I), ie. Vestiti.

e. Veil of filaments only—C. filamentifer, C. luteocephalus, etc., ie. Impexi. Fig. G.

f. Mixture of thin- and thick-walled, often brown, elongate and rounded cells—C. heptemerus & C. curtus, Fig. J.

APPENDIX I

AGARIC-GASTROMYCETOID RELATIONSHIPS: (ILLUSTRATED ON PLATE 14; FIGS. A-N)

TABLE 1 GENERAL RELATIONSHIPS

Agaricaceae
                                        Gyrophragmium delilei
                                        Endoptychum pro parte
                                        Smithiogaster volvoagaricus

Amanitaceae
                                        Torrendia pulchella Fig. H

Bolbitiaceae
    Agrocybe
                                        Cytarrophyllum besseyi
    Bolbitius
                                        Gastrocybe lateritia
                                        Tubariopsis torquipes

Coprinaceae
    Coprinus
                                        Montagnea arenaria
                                        Longula texensis
    Panaeolus
                                        Panaeolopsis sanmartianaiana

Cortinariaceae
    Descolea
                                        Setchelliogaster tenuipes
    Pholiota
                                        Nivatogastrium nubigenum

Entolomataceae
                                        Rhodogaster chilensis Fig. J.
                                        Richoniella leptonispora Fig. I

Gomphidiaceae
    Chroogomphus
                                        Brauniellula nancyae
    Gomphidius
                                        Gomphogaster leucosarx

Lepiotaceae
    Lepiota
                                        Endolepiotula muzealii
                                        Notholepiota areolata
                                        Secotium gueinzii
                                        Phyllogaster pholiotoides
    Macrolepiota
                                        Neosecotium macrosporum
                                        Endoptychum agaricoides

Paxillaceae
    Paxillus
                                        Austrogaster marthae

    Phylloporus   *Paxillogaster luteum* Fig. D.
                                                     Singeromyces ferrugineum

Pluteaceae   *Brauniella (Brauniellaceae)* Fig. C.

Strophariaceae
    Psilocybe   *Clavatogaster virescens*
                                                Cytarrophyllopsis cordispora

    Stropharia   *Weraroa erythrocephalum*, *W. Coprophila* Fig. A.
                                               Galeropsis desertorum, *G. Polytrichoides* Fig. B.
                                               Tympanella galanthina

Thricholomataceae
    Laccaria
       (*Hydnangiaceae*)   *Hydnangium carneum*

    Fayodia/Oudemansiella   *Gigasperma cryptica*

                             TABLE 2   SELECTED GROUPS

                                 <u>a</u>)   <u>Agaricales</u>

Cortinarius
        gasteroid agarics: *C. bigelowii*, *C. magnivelatus*, *C. velatus*,
         & *C. wiebeae*

                                         Thaxterogaster (T. pinque) Fig. K.
                                             s.g. *Thaxterogaster*
    sg. *Myxacium*   *T. magellanicum*

    sg. *Phlegmacium*   *T. pisciodorum*
                                            sg. *Hemicortinarius*

    sg. *Telamonia/Hydrocybe*   *T. ochraceoaureum*
                                             T. cretaceum
                                             T. cartilagineum

    sg. *Sericeocybe/Dermocybe*   *T. albocanum*
                                             sg. *Volvigerum*

    sg. ??   *T. conei*

b) Russulales

Russula
   hypogeous agarics: *R. brevipes, R. foetens*

                                       Macowanites spp. Fig. L.
                                            Martellia pro parte Fig. F.
                                            Elasmomyces Fig. G.
                                            Gymnomyces pro parte; Cystangium

Lactarius
                                            Arcangeliella
                                            Zelleromyces

c) Boletales

Suillus	*Castroboletus*
semi-hypogeous	
S. riparius	*G. imbellus*
	G. suilloides
S. decipiens	*G. laricinus*
Tylopilus	
semi-hypogeous	
T. humilis	*G. boedijnii*
Leccinum	*G. scabrosus* Fig. M.
Boletus	
sg. *Boletus*	
sect. *Luridi*	*G. turbinatus* Fig. N.
sect. *Edulis*	*G. subalpinus*
sg. *Xerocomus*	*G. xerocomoides* - ? *Gymnopaxillus*
	(see Paxillaceae)

APPENDIX II

AGARICOID-CYPHELLOID RELATIONSHIPS
(ILLUSTRATED ON PLATE 15; FIGS, A-P)

almost poroid: parallel poroid life-forms are found in *Dictyopanus* (Panellaceae) and *'Poromycena'* (Myceneae)

TABLE 1 TRUE AGARIC RELATIONSHIPS

Tricholmataceae
    Clitocybe (Clitocybineae)   *Hypsizygus*
                                       (pleurotoid)

        - Clitocybeae   *Calyptella*

    Crinipellis Marasmineae   *Chaetocalathus*
       (Crinipellineae   (pleurotid)   *Flagelloscypha* Fig. A&B.
                                                                       Lachnella

    Marasmiellus (Collybieae)   *Cheimonophyllum*
                                             (pleurotoid)   *Campanella*
                                                                         Skepperiella Fig. H.
                                                                         Mniopetalum

    Marasmius (Marasmineae)
      (Marasmieae
         sect. Epiphyllii   *Gloiocephala* Fig. F&G.
         sect. Hygrometrici   *Manurpia*
         sect. Alliacei   *Hymenogloea*

    Mycena (Myceneae)   *Cellypha* Fig. E.

    Omphalina (Tricholomatineae)
                              O. rickenii
                              Leptoglossum   *Phaeotellus* Fig. M.
                              Arrhenia Fig. I   *Cyphellostereum* Fig.

        (Resupinateae)   *Resupinatus*
                                       (pleurotoid)   *Stigmatolemma**
                                                                        Stromatoscypha

        (Panelleae)   *Panellus* etc. Fig. O, P & P.
                                         (pleurotoid)

        (Rhodoteae)   *Rhodotus palmatus*
                                         (pleurotoid)

Crepidotaceae
    Crepidotus　　　　　　　*Pleurotellus-*
                             graminicola
                     Chromocyphella Fig.K
        C. antillarum　　　　　　　　　　　*Cyphellopsis** Fig. C&L
           grp.　　　　　　　　　　　　　　*Episphaeria*
                                     Merismoides*
                                       Pellidiscus
                                       Phaeosolenia*

TABLE 2 AGARICOID LIST

Non-agaric cyphellaceous fungi

Aleurodiscus	Corticiaceae
Auriculariopsis	Corticiaceae
Cytidia	Corticiaceae
Favolaschia	Favolaschiaceae Fig. J.
Heteroscypha	Tremellaceae (Tremellales)
Henningscomyces	Schizophyllaceae Fig. D.
Mycobonia	Polyporaceae
Papyrodiscus	Polyporaceae near *Flavodon* questionably related to Tricholomataceae
Physalacria	
Porotheleum	Schizophyllaceae
Tremelloscypha	Tremellaceae (Tremallales) near *Tremiscus*

APPENDIX III

GLOSSARY

Agaricales — Order of fleshy, putrescent basidiomycetes with spores produced either on plates, or in tubes, borne beneath a well-defined cap or pileus; now split into four orders: Agaricales, Pluteales, Tricholomatales and Boletales: Covers those fungi which collectively are called mushrooms and toadstools, and gives rise to the common word agaric.

Allocyst — thin- or slightly thick-walled, swollen cell formed in culture and acting as a repository; not a propagule.

Anamorphic — the asexual stage of a fungus with a known sexual stage; the imperfect state.

Angiocarpic — with the hymenium developing and achieving maturity within a cavity (or cavities) in the basidiome.

Antibiotic — a chemical substance produced by an organism which inhibits or destroys bacteria or other micro-organisms.

Apobasidium — a basidium on which the meiospores are borne asymmetrically and from which they are forceably discharged.

Arthroconidium — largely undifferentiated mitospore formed seriately by fragmentation of the hypha.

Ascomycetous — having the characters of an Ascomycete (member of the Ascomycotina), ie. the possession of an enveloping sac (ascus) containing meiospores.

Autolysis — disintegration of cells or tissue by the breakdown of the cells themselves as in *Coprinus*.

Basidiome — the entire structure in basidiomycetes bearing the reproductive tissue; variously known as basidiocarp, sporocarp, sporophore, fruiting body and fruit-body.

Basidiomycotina — the major taxonomic group containing those fungi possessing basidia; gives rise to the common group name basidiomycete (-s).

Basidiospore — the meiospore produced at the apex of a basidium often referred to simply as the spore.

Basidium   the cell in the basidiomycetes in which fusion followed by reproductive division (meiosis) takes place; in the agarics, this is a single cell.

Bipolar   with sexual factors of two kinds.

Blocker   the possession of a gene or group of genes which prohibit fusion of the hyphae and therefore transference of nuclei to form normal dicaryons.

Bromatia   massing together of rounded, swollen, thin-walled hyphal-ends to form small aggregates; used by ants as food.

Bulbil   restricted to a small sclerotium formed from a few layers of cells; in the forest pathology used erroneously for the curling and twisting of vegetative hyphae around each other to form small knots; used erroneously for anamorphic clusters of cells in *Coprinus clastophyllus* which are now termed catervae.

Caryotic
   caryo-, cary-, carya-, combining form meaning nucleus: see Karyotic, mono-; di-; hyphal cells containing single nuclei or two nuclei of different mating types respectively.

Casing soil   the sterile or partially sterile soil placed on cultures, whether in commerce or in the laboratory.

Cephalosus   the diapsore of *Mycena citriicolor*; erroneously called a gemma—the whole structure bearing the cephalosus has erroneously been called stilboid and placed in the genus *Stilbum*.

Chlamydospore   a thick-walled secondary spore developed from hyphae, but not borne on basidia or conidiophores.

Chrysocystidium (-ia) sterile cell containing an amorphous body with affinity for cotton blue and becoming yellow in ammoniacal solutions.

Clamp-connection a small semicircular hollow protuberance laterally, attached to the walls of two adjoining cells and arching over the septum which separates them; often contracted to clamp (= bouclee). Hyphae possessing such structures are said to be clamped or nodose-septate.

Clavaria-like   resembling a member of the genus *Clavaria* in its traditional sense; either coral- or fairy- club-shaped—*clavarioid*.

Conidial of or pertaining to conidia; bearing conidia, characterised by forming or bearing conidia; conidi-, or conidio-, appertaining to conidia - conidium (-ia) a mitospore, ie. neither produced by meiosis (caryogamy) nor produced in a sporangium.

Conidiogenesis the process of producing conidia.

Coprophilous growing on dung (≡ fimicolous).

Common - A,....B the mating factors in a heterothallic fungus which lead on confrontation to the formation of a dicaryon.

Cyphelloid having the hymenium developed in a cup-shaped basidiome.

Cystidium sterile, differentiated cell produced either in or on the hymenium or the sterile surface of the basidiome.

De-dicaryotization the process of segregating the two mating types in a dicaryotic mycelium.

Diaspore an ecological term for the dispersive stage; includes asexual propagules, eg.conidia, bulbils, sclerotia, and sexual propagules, eg.basidiospores.

Dicaryotic possessing two independent haploid nuclei (binucleate) in the majority of the vegetative hyphal cells, ie.containing two nuclei with different mating types (=dicaryotic diploid cell): dicaryon—the binucleate condition of the fungus: dicaryophase —that phase of the fungus characterised by a dicaryon: dicaryotization the process of forming a vegetative stage derived from two parents of opposing mating types.

Dikaryotic = dicaryotic

Diverticulate having short, vertical branchlets (diverticulae) appearing at more or less right angles.

Facial cystidium sterile hymenial cell formed on the gill-face or within the tube (= pleurocystidium).

Fungi imperfecti general term for all those form-genera and form-species based on asexual diasporic stages; only some form-species have teleomorphs. Placed in the Deuteromycotina.

Gasteroid (or gastroid) having the hymenium enclosed.

Genotypic the expression of the fundamental gene construction; opposite

Gloeocystidium (or gleo-) a sterile cystidium containing oily or resinous to distinctly granular contents with affinity for cotton blue stain.

Gymnocarpic   with the hymenium developing exposed to the environment throughout formation.

Hemiangiocarpic with the hymenium developing within a cavity in basidiome but soon becoming exposed to the environment and completing development exposed.

Holobasidium   a cell in which meiosis is completed but not followed by septal formation.

Homobasidiomycete a fungus referrable to a group of basidiomycetes possessing non-septate basidia.

Homothallic   formation of fertile basidiomes from a single nuclear type.

Humicolous   growing on partially decomposed leaves and similar soft, rotting vegetation.

Hymenium   the fruiting surface or spore-bearing layer of the basidiomycetes; consists of a continuous layer (palisade) of basidia and analogous cells and gives rise to the adjective hymenial, ie. tissue immediately bearing the hymenium and subhymenium, eg. gills, tubes: hymenophoral trama—tissue supporting the hymenium.

Hypha   a fungus filament; hyphal—pertaining to hyphae.

Incompatible   not cross-fertile when mater together; sexually uncongenial.

Karyotic   Kary-, Karya-, Karyo-: a combining form meaning nucleus: preferred by many to caryo-; see mono- and dicaryotic.

Laticiferous   of hyphae yielding a milk-like fluid on injury or visible under the microscope as hyphae with milky, granular contents; same as lactiferous; lactifer—a latex-bearing hypha or hyphal element.

Lignicolous   growing in or on trunks, stumps, shanks, branches or twigs.

Locus   the operative position of a gene on a chromosome (the fundamental thread or chair of proteins composing dividing nuclei).

Luminescence   that phenonemon which emits light energy; bio—the phenonemon applied to organisms.

Marginal cystidia   sterile hymenial cells formed at the gill-margin or at the orifice of tubes (= cheilocystidium).

Mat (culture)   the aerial vegetative mycelium formed by fungi in culture.

Meiospore   spore formed after fusion and reproduction division (meiosis).

Mites   small arachnids which feed on basidiomes and vegetative stages of agarics and other fungi, both in the field and laboratory.

Mitospore   spore formed after asexual (mitoric) division of the nucleus.

Mycelium (-ia)   collective term for a group or mass of fungus filaments (hyphae); mycelial—pertaining to the mycelium or having properties of the mycelium.

Mycorrhizal   a range of intimate or loose symbiotic associations of fungi with short roots of vascular plants; mycorrhizal—pertaining to mycorrhiza (fungus-root).

Nuclear migration   movement of nuclei through the hyphae from one part of the culture to the other; usually refers to the movement of the dicaryotizing nucleus after confrontation and fusion of two mating types.

Oidium   meristem-arthroconidia of Erysiphales. Erroneously used for thin-walled thallic arthroconidia of basidiomycetes.

Paarige branching   unusual formation of clamp-connections on several closely situated septa to form a small knot.

Parasite   an organism living upon another living organism and deriving food from it; parasitic—living in or on another organism.

Phosphorescence   the process by which organisms produce light energy (yellow or blue-green) involving the activity of ATP (adenosine tri-phosphate).

Pileate   possessing a cap (or pileus).

Pleurotoid   having one or more characters of the genus Pleurotus; usually applied to a gilled fungus with lateral or excentric stipe, or lack of same.

Polyporales   the order of tough, woody basidiomycetes with indefinite growing margin or amphigenous hymenium; includes the poroid fungi (except boletes), fairy-clubs, hedgehog fungi, etc., Replaces the name Aphyllophorales.

Polysaccharide sugar composed of chains of simple -sugar units.

Primordium (-ia) the initial group of hyphae which develops into the basidiome; primordial initial.

Propagule   applicable to either sexual or asexual dispersive stages of a basidiomycete or other fungus.

Refractive bodies amorphous or irregular bodies with high optical refraction found in culture; either within or external to the cell. Hyphal walls are often described as refractive because of their glassy appearance under the microscope.

Rhizoid   finely branched, root-like, radiating mycelia.

Rhizomorph   a distinctive cord or strand of compacted mycelia.

Root (-cortex; -stele) the underground portion of angiosperms and gymnosperms, the rootlets being easily separable under the dissecting microscope into the outer zone (cortex) and the vascular medulla (stele).

Saprophyte   an organism living on dead organic matter.

Sclerotium (-ia) a resting body composed of massed hyphae with (pseudosclerotium) or without host tissue or soil and usually with a darkened rind.

Spermatium (-ia) non-motile gamete (male) which unites with a receptive hypha (female); erroneously applied as a synonym of oidium. Some thallic arthnoconidia may act as psermatia.

Spore-print   the pattern produced by the forceably ejected basidiospores when a cap is placed gills down on a piece of paper.

Statismospore spores produced on the (statismo-) basidium of a gasteromycete; not forceably discharged.

Sterile   a substrate deprived of organisms; rid of microbes.

Stipitate   possessing a stem (or stipe).

<u>Teleomorph</u>	the sexual stage of a fungus.
<u>Tetrapolar</u>	with sexual factors segregated into four groups.
<u>Thallus</u> (<u>i</u>)	the general term for the assimilative phase of a basidiomycete; <u>thallic</u>—pertaining to the thallus; <u>homo</u>—the formation of zygotes or dicaryotic cells from a single thallus; <u>hetero</u>—the formation of zygotes or dicaryotic cells from two different parent thalli.
<u>Trama</u>	a (pseudo-) tissue supporting the reproductive cells; <u>hymenophoral trama</u>—the tissue between adjacent hymenia; <u>pileus-trama</u> fleshy portion of the pileus.
<u>Veil</u>	a protective structure derived either in hemiangiocarpic basidiomycetes from the original tissue or formed secondarily in some initially gymnocarpic species; <u>velar</u>—pertaining to a veil.
<u>Velangiocarpic</u>	with the hymenium developing within the protection of a membrane which later disrupts and/or disintegrates.

APPENDIX IV

REFERENCES AND ADDITIONAL READING

This list of references is intended not to be extensive but to act as a guide allowing users of the text to follow up the techniques outlined.

I. GENERAL TEXTS

Buller, A.H.R. (1909-1934). *Research in Fungi* I-VI, Longmans, London
Booth, C. (1971). *Methods in Microbiology* Vol IV, Academic Press, London
Commonwealth Mycol. Inst. (1968). *Plant Pathologists Handbook*, Kew, England.
Dade, H.A. & Gunnell, J. (1966). *Class work with Fungi*, Commonwealth Mycol. Inst., Kew, England.
Hawksworth, D.L. (1974). *Mycologist's Handbook*, Commonwealth Mycol. Inst., Kew, England.
Stevens, R. ed. (1974). *Mycology Guide Book* Guidebook Committee, Mycological Society of America, Washington Univ. Press, Seattle.

II. MEDIA

Badcock, E. C. (1941). *Trans. Brit. Mycol. Soc.* 25; 200-205—Saw-dust agar
Badcock, E. C. (1943). *Trans. Brit. Mycol. Soc.* 26; 127-32—Saw-dust agar
Berliner, M.D. (1961). *Mycologia* 53; 84-90—Bread agar
Watling, R. (1963). *Nature*, London 197, 717-718—Paper-pulp agar

III. TECHNIQUES

A. COLLECTING

Largent, D. (1973). *How to Identify Mushrooms to Genus I* Macroscopic Features, Mad River Press, Eureka, California.
Largent, D. Johnson, D. & Watling, R. (1977). *How to Identify Mushrooms to Genus III*: Microscopic features, Mad River Press, Eureka, California.
Savile, D.P.O. (1962). *Collection and care of Botanical Specimens*, Can. Dept. of Agric., Ottawa.
Taylor, J. (1979). *Amateur Gardening*, June 23, 1979.
Watling, R. (1969). *Microscopy* 31; 95-105.
Watling, R. (1973). *Identification of Larger Fungi*, Hulton Educ. Press, Amersham, England.

B. LABORATORY PROCEDURES

1. *General* (also see I above)

Lloyd, A.O. (1965). *Int. Biodet. Bull* I: 10-13 (mounting with adhesive tape)
Omar, M.B., Bollard, L. & Heather, W.A. *Bull.Brit. Mycol. Soc.* 13:31-32—polyvinyl alcohol

2. <u>Techniques for securing inoculum</u>

   a. Isolation from tissue

   Gallymore, B. (1949). *Trans. Brit. Mycol. Soc.* 32; 315-317.
   Harley, J.L. & Ward, J.S. (1913). *Trans. Brit. Mycol. Soc.* 38; 104-118.
   Watling, R. (1971). *Methods in Microbiology* IV, Isolation...... Basidiomycetes; Homobasidiomtcetidae, Academic Press, London.

   b. Isolation from soil etc.

   Nokrans, B. (1949). *Svensk Bot. Tidskr.* 43; 485-490.
   Cartwright & Findlay, W.P.K. (1958). *Decay of timber and its prevention.* Her Maj. Stat. Office, London.
   Warcup, J. (1957). *Trans. Brit. Mycol. Soc.* 40; 237-264.
   Warcup, J. (1959). *Trans. Brit. Mycol. Soc.* 42; 427-435.
   Warcup, J. & Talbot, P.H.B. (1962). *Trans. Brit. Mycol. Soc.* 45; 495-518.
   Warcup, J. & Talbot, P.H.B. (1966). *Trans. Brit. Mycol. Soc.* 49; 427-438.

   c. Spore germination

   Fries, N. (1941). *Archiv für Mikrobiologie*, 12: 266-284.
   Fries, N. (1966). in Madelin, *The Fungus Spore*, Butterworths, London.
   Fries, N. (1977). *Mycologia* 59; 843-850.
   Fries, N. (1978). *Trans. Brit. Mycol. Soc.*, 70; 319-324.
   Gottlieb, D. (1950). *Bot. Rev.* 16; 229-257.
   Losel, D.M. (1964). *Ann. Bot.* (N.S.) 28; 465-478.
   Losel, D.M. (1967). *Ann. Bot.* (N.S.) 31; 417-425.
   Madelin, M.F. (1966). *The Fungus Spore*, Colston Papers, No. 18. Butterworths, London.
   Peterson, R. (1960). *Mycologia* 52; 513-514.
   Robbins, W.J. (1950). *Mycologia* 42; 470-476.
   Watling, R. (1963). *Nature*, London 197; 717-718.

3 & 4. <u>Cultivation and Maintenance</u>

   Martin, J.P. (1950). *Soil Sc.* 69; 215-232.
   Melin, E. (1953). *Am. Rev. Pl. Physiology.* 4; 325-346.

Nokrans, B. (1949). *Svensk. Bot. Tidskr.* 43; 485-490.
Russell, P. (1956). *Nature*, London. 177: 1038-1039.

5. *Techniques for inducing fructification*

Bigg, W. & Alexander, I (in press). A culture unit for the study of nutrient uptake by nycorrhizal plants under aseptic conditions. *Soil Biol. Biochem.*
Dawson, J.R. & Johnson, R.A., Adams, R. & Last, F.T. (1965). *Ann. Appl. Biol.* 56; 243-284.
Doyle, P., Morrison, R. & Whalley, A.J. (1976). *Trans Brit. Mycol. Soc.* 66; 173-174 (use of Hexahydro 1,3,5 tris (2 hydroxethyl 1)-s-razine (Grotan).
Flentje, N.T. (1957). *Trans. Brit. Mycol. Soc.* 40; 322-336.
Hayes, W.A. (1972). *Mushroom Science,* 8; 663-674.
Melin, E. (1948). *Trans. Brit. Mycol. Soc.* 30; 92-99.
Plunkett, B.E. (1956). *Ann. Bot.* 20; 563-586.
Takemaru, T. & Kamada, T. (1969). *Rept. Tottori Mycol. Inst.* (Japan) 7; 127-140. A bibliographic review.
Volz, P.A. & Beneke, E.S. (1969). *Mycopath & Myco app.* 37: 225-253.
Warcup, J.H. & Talbot, W.P. (1962). *Trans. Brit. Mycol. Soc.* 45: 495-518.

C. STAINING AND SECTIONING TECHNIQUES

Bennell, A., Christopher, P., & Watling, R., (1978). *Trans. Brit. Mycol. Soc.* 71: 512-515 (ultrathin sections for light and electron microscopes).
Disbrey, B. & Watling, R. (1967). *Mycol. Myco. appl.* 32: 81-114 (Staining paraffin sections).
Watling, R. & Nicoll, H., (in press). *Trans. Brit. Mycol. Soc.* (temporary sections to study development).

D. GENETIC TECHNIQUES

Kemp, R.F.O. (1971). *Trans. Brit. Mycol. Soc.* 55: 493-496 (micromanipulator).
Seaby, D.A. (1977). *Bull. Brit. Mycol. Soc.* 11: 52-54 (single spore isolation).

IV. CHARACTERISTICS OF AGARICS IN CULTURE

A. INTRODUCTION

Burnett, J. H. (1968). *Fundamentals of mycology,* Arnold, London.
Kühner, R. (1977). Les grandes lignes de la Classification des Boletales, *Bull. Soc. Linn Lyon* 46: 13-81.

Kuhner, R. (1977-9). Les Grandes lignes de la classification des Agaricales, Pluteales, Tricholomatales, Bull. Soc. Linn. Lyon. 46; 81...48: 593. continuing.
Singer, R. (1975). The Agaricales in Modern Taxonomy, J. Cramer Vaduz.
Watling, R. (1978). From Infancy to Adolescence—Suppl. Trans. Bot. Soc. Edin. 42; 6-73.

B. DEVELOPMENT OF BASIDIOME

Reijnders, A.F.M. (1963). Les Problemes du development des carpophores Agaricales et de quelques groupes voisins, Junk, Hague.
Watling, R. (1975). Studies in Fruit-body development in the Bolbitiaceae and the implications of such work in studies in Higher Fungi, Beih, Nova Hedw. 51: 319-346.
Watling, R. (1971). Persoonia 6: 281-289.

C. SECONDARY SPORES

Kendrick, W.B. & Watling, R., (1979) in Kendrick. The Whole Fungus, Mitospores in Basidiomycetes 473-546, 2nd Int. Mycol Conference (1977) Can. Nat. Mus. Sc. Ottawa and Kananaskis Foundation, Canada.
Watling, R. (1978). The morphology, variation and ecological significance of anamorphs in the Agaricales, as above 453-472.
Watling, R. (1980). Taxonomy of Fungi; Proc. Int. Symp on Taxonomy of Fungi Part II Madras.

D. ADDITIONAL CULTURAL CHARACTERS

Campbell, W.A. (1938). Bull Torrey Bot. Club, 65; 31-78.
Corner, E.J. H.(1932). Ann. Bot. (London). 46;71-111.
Fritz, C. (1923). Trans. Toy. Soc. Can., ser. 5, 17: 191-288.
McKeen, C.G. (1952). Can J. Bot. 30: 764-787.
Maxwell, M.B. (1954). Can. J. Bot. 32; 359-280.
Miller, O.K. (1971). in Petersen, Evolution in the Higher Fungi, The relationship of cultural characters to the taxonomy of the agarics. Univ. of Tenn. Knoxville.
Nobles, M. (1948). Can. J. Res. C. 36; 281-431.
Nobles, M. (1965). Can. J. Bot. 43:. 1097-1139.
Refshauge, L.D. & Proctor, E.M. (1936). Proc. Royal Soc. Victoria 48, (NS): 105-123.
Stalpers, J.S. (1978). Identification of wood-inhabiting Aphyllophorales in pure culture. Studies in Mycology No. 16 Baarn, Netherlands Specific to boletes.
Pantidou, M. (1961). Can J. Bot., 39; 1149-1162.
Pantidou, M. & Groves, J.W. (1966). Can J. Bot. 44: 1371-1392.
Pantidou, M. & Watling, R. (1970). Notes Roy. Bot. Gdn. Edin., 30: 207-237. Bioluminescence.

Berliner, M.D. (1961). Mycologia 53; 84-90. See also Macrae under 5b.

V. EXPERIMENTS, STUDIES, & TESTS

A. CHEMICAL TESTS

General: Documentation of Threshold limit values (1977) for substances in workroom Air (1978). Sec. Trans. Americ. Conf. Govt. Industrial Hygiene, Cincinnati, Ohio.

1. Cultures

Davidson, R. W., Campbell, W.A. & Blaisdell, D.J. (1938) J. Agric. Res., 57: 683-95.
Kaarik, A. (1965). Studia Forestalia Suecica 31: 1-80.
Taylor, J.B. (1974). Ann. Applied Biology 78; 113-123.
Taylor, J.B. (1977). Ann. Applied Biology 85; 181-193.
Bavendamm, W. (1928). Z. Pfl. Krank.Pfl.Schutz, 38; 257-276.

2. Basidiome material

Watling, R. (1971). in Methods in Microbiology IV: Chemical Test in Agaricology. Academic Press, London.

B. GENETIC STUDIES

Day, P. (1959). Heredity, 13; 81-88—*Coprinus cinereus* as *C. lagopus*.
Kemp, R.F.O. (1970). Trans. Brit. Mycol. Soc. 54; 488-489.
Kemp, R.F.O. (1974). Trans. Brit. Mycol. Soc. 62; 547-555.
Kemp. R.F.O. (1975). Trans. Brit. Mycol. Soc. 65; 375-388.
Kniep, H. (1928). Die Sexualitat der niederen Pflanzen, Fischer, Jena.
Lange, M. (1952). Dansk. Bot. Ark. 14; 1-64.
Moore, D. (1966). Nature, London 209; 1157-1158.
Moore, D. (1979). New Phytol 83; 695-722—*Coprinus cinereus*.
More, K., Fukai, S., & Zennoyozi, A. (1976). Mushroom Science 9; 391-404.
Papazain, H.P. (1950). Bot.Gaz. 112; 138-139.
Rosinski, M.A. & Robinson, A.D. (1968). Am.J. Bot. 55; 242-246—*Panus tigrinus/Lentodium squamulosum*.
Macrae, R. (1942). Can. J. Res., 20; 411-434—*Panus stipticus*. Effects of light.
Miller, O.K. (1967). Can J. Bot. 45; 1939-1943—*Panus fragilis*.
Plunket, B.E. (1958). Mushroom Grower's Assoc. Bull. 99; 76-79.

Armillaria studies:

Anderson, J.B. & Ulbrich, R.C. (1979). Mycologia 71; 402-414.

Korhonen, K. (1978). Karstenia 18; 31-42.
Ulbrich, R.C., & Anderson, J.B. (1978). J. Exp. Mycol. 2; 119-129.

VI. SUGGESTED EXERCISES

A. LABORATORY EXPERIMENTATION

1. General

a. Vegetative stage

Fries, N. (1949). Svensk. Bot. Tidskr, 43; 316-342.
Modess, O. (1941). Symb. Bot. Upsal. 5; 1-147.
Nokrans, B. (1949). Svensk. Bot. Tidskr. 43; 485-490.

b. Fruiting

Billie-Hansen, E., (1953). Bot. Tidskr., 50; 81085.
Borriss, H. (1934). Planta 22; 28-69.
Lu, D.C., (1965). Am. J. Bot. 52; 432-437.
Plunkett, B.E. (1956). Ann. Bot. 20; 63-586.
Manachere, G. (1976). Mushroom Science, 9; 783-798.
Hagimoto, H. & Konishi, M. (1959). Bot. Mag.(Tokoyo) 72; 359-366: ibid 73: 283-287.
Hagimoto, H. (1963). Bot. Mag. (Tokoyo) 76: 256-263.

c. Cultivation

Chang, S.T. & Hayes, W.A. (1978). Biology and Cultivation of Edible Mushrooms, Academic Press, London.

Singer, R. (1961). Mushrooms and Truffles - Botany, Cultivation & Utilization. Leonard Hall, London. Also see Mushroom Grower's Journal, Mushroom Science, etc.

2. Genera

a. *Agaricus*

Atkins, F.C. (1956) Mushroom growing today, Faber & Faber, London.
Bonner, J.T., Kane, K.K. & Levy, R. H. (1956). Mycologia 48; 13-19.
Duggar, B.M. (1905). Bull. U.S. Dept. Agric. 85; 1-60.
Jackson, R.C. O. (1951). Mushroom Growing—a practical manual, English Univ. Press, London.
Raper, C. & Kaye, G. (1978). Sexual and other relationships in

the genus *Agaricus*. Journ. Gen. Microbiol., 105; 135-151.
Physiology
Treschow, C. (1944). Dansk. Bot. Akiv. II(6): 1-180.
Mader, E.O. (1943). Phytopathology, 43; 1134-1145.

b. *Coprinus* also see VIII.

Anderson, G.E. (1971). The Life History of Genetics of *Coprinus lagopus*.
Harris Biol. Supplies, Weston-super-Mare.
Brunswick. H. (1924). Bot. Abhandlungen, Herausgeg v. Goebel. 5; 1-152.
Kemp, R.F.O. (1975). Trans. Brit. Mycol. Soc., 65; 375-388.
Lange, M. (1952). Dansk. Bot. Ark. 14; 1-64.

c. *Stropharia*

Puschal, J. (1969). Deutsche Gartnerpost, 42; 10-15.

d. *Psilocybe*

Stamets, P. (1978). Psilocybe mushrooms and their allies, Homestad Book Co., Seattle.
Pollock, S. (1977). Magic Mushroom cultivation, Herbal Med. Res. Foundation, Texas.

e. *Pholiota*

Deneyer, W.B.G. (1960). Can. J. Bot. 38; 909-920.
Farr, E.R., Miller, O.K., & Farr, D.F. (1977). Can J. Bot. 55; 1167-1180.

f. *Agrocybe*

Cailleus, R. & Diop, A. (1976). Mushroom Science. 9; 607-619.

g. *Volvariella*

Chua, S.E. (1976). Mushroom Science, 9; 701-706.
Chua, S.E. & Ho, S.Y. (1973). World Crops, 25; 90-91
San Antonio, J.P. & Fordyce, C. (1972), Hort. Sci. 7; 461-464.

h. *Flammulina*

Gruen, H.E. (1969). Mycologia 61; 149-166.
Gruen, H.E. & Sheue-Ling, Wm. (1972). Mycologia 64; 995-1007.
Kinugawa, K. & Furukawa, H. (1965). Bot. Mag. (Tokoyo) 78;

240-244.

i. *Pleurotus*

Bresinsky, A., Hilber, O., & Molitoris, H.P., (1977). The genus *Pleurotus* as an aid for understanding the concept of species in Basidiomycetes, in Species concept in Hymenomycetes, J. Cramer, Vaduz.
Eugenio, C.P. & Anderson, N.A. (1968). Mycologia 60; 627-634.
Eger, G., Gottwald, H.D., & Netzer, U.von. (1976). Mushroom Science 9; 567-573 also 578-583.
Cailleux, R., Diop, A. & Macaya-Lizano, A. (1976). Mushroom Science 9; 595-606.

j. *Lentinula*

Ando, M. (1976). Mushroom Science 9; 415-422.

k. *Schizophyllum*

Wessels, J.G.H. (1965). Wentia 13; 1-113 (good review).
Raper, J.R. & Krongelb, G.S., (1958). Mycologia 50; 707-740.

B. LABORATORY ECOLOGY

Richardson, M.J., & Watling, R. (1971). Keys to Fungi on Dung, Suppl. Bull. Brit. Mycol. Soc., 2; 18-43 (1968) and 3; 86-88; 121-124 (1969).
Additional
Cooke, Wm. Bridge (1968). Mycopath & Mycol. app. 34; 305-316.

APPENDIX V

CHECK LIST OF CHEMICALS AND EQUIPMENT

General apparatus

- Aluminium foil (cooking foil)
- Blue tac
- Bunsen burner
- Cellophane paper (or sausage skin)
- Cellulose enamel (or comm. cover-slip sealer)
- Chinagraph pencil (or magic marker)
- Cigarette papers
- Cooker: domestic, gas or electric or hot air oven)
- Cotton wool (or absorbent wool)
- Double saucepan (or water bath)
- Enclosed steamer
- Fairly coarse sand (garden sand without additives)
- Filter papers (or blotter paper)
- Funnel (glass or plastic)
- Incubator
- Indian ink
- Mohr clip
- Nail varnish
- Paper-pulp
- Panchromatic film
- Paraffin wax
- Peat (garden)
- Pressure cooker (or autoclave)
- Refrigerator (domestic)
- Retort stand (or equivalent with clamped ring
- Scotch tape
- Sterile water (distilled)
- Spirit lamp (or Bunsen burner)
- Sterilised soil
- Thermometer
- Tubing (plastic or rubber)
- Vaseline (=Petroleum jelly)
- Vermiculite

Glass-ware

- Boiling tubes 1 inch
- Casseroles (or sandwich boxes)
- Flasks (flat bottom; or equivalent) 1 litre
- Glass plate (sheet suitable for swabbing down)
- Jars (squat domestic jars with screw-caps)
- McCartney phials (or equivalent) 1 oz.
- Medical flats (or equivalent) 4 oz.
- Petri dishes (or polystyrene pre-sterilised disposable Petri dishes)
- Tubes (thick-walled, pyrex)

Microscope equipment

- Biscuit (or cake)-cutter objective
- Binocular viewer, or binoc. microscope-dissecting microscope
- Compound microscope with x 10 and x 40 objectives: x 100 objective optional x 7 or 10 eyepieces
- Glass rods (small lengths: glass rings
- Glass slides and cover-slips
- Micro-pipette (or tube with bulb, or syringe)
- Needles with points, with loop and nichrome end
- Tweezers
- Welled slides

Collecting apparatus

- Basket
- Brown paper-bags
- Colour chart, see pg.
- Fern trowel or fork

Collecting apparatus cont'd.

> Selection of plastic tubes and tins
> Silica gel (preferably fairly coarse grade)
> Small drier (either hand made with bulb as heater or commercial drier)
> Strong knife (sheath or kitchen)
> Waxed paper

Chemicals

> activated charcoal
> agar-agar
> alcohol (ethanol—comm. grade)
> benomyl 'benlate'
> σ-camphor
> chloros solution (comm. grade)
> copper sulphate
> formalin (formaldehyde; comm. grade)
> furfural (furfuraldehyde)
> Hydrochloric acid (concentrated; comm. grade)
> gallic acid (or tannic acid)
> gelatin
> glacial acetic acid
> glycerine (glycerol; comm. grade)
> gum chloral,

Chemicals cont'd.

> iodine crystals
> mercuric chloride
> methyl benzoate
> methylated spirits (methanol) (comm. grade)
> mineral oil
> naphthalene
> napthol
> neomycin sulphate
> pancreatin
> para-dichlorobenzene
> σ phenyl-phenol
> potassium tellurite
> propulene oxide
> rose bengal
> saline solution
> specific chemicals

Specific chemicals

> antibiotics
> chemical tests
> media recipes
> sectioning recipes
> stain recipes
> steptomycin
> tannic acid
> teepol (and household detergent)
> thiamine
> thymol

All the materials above are available 'over the counter' from any biological suppliers. In the text, alternatives have usually been offered as far as containers, glass-ware etc. are concerned, in order to allow the amateur to enjoy the study of larger fungi in the laboratory. Some of the more sophisticated techniques require compounds which must be purchased from a chemical supplier, although the majority of chemicals can be obtained from or even made up by a local pharmacist. Alternatives are also offered in the text for several of the techniques adopted. However, the most useful and powerful tools at present for studying populations of agarics are the tissue-grinder and some glycerol solutions, and a microscope (see pg. 43).

APPENDIX VI

ABBREVIATIONS

cm	centimetre (s)
comm. grad.	Commercial grade in respect to bulk chemicals
g	grams
Greek letters	used in chemical nomenclature
	alpha α
	ortho-
	para- p
hrs.	hours
l.	litre (1000ml)
L.S.	Longitudinal section
M	Molar in molar solutions equivelent to the molecular weight
mg.	millegram = 1/1000g.
mins.	minutes
ml.	millilitre = 1/1000 l.
mm	millemetre (s)
N	Normal as in normal solutions. Equivalent weight of compound in 1000 ml.
sec.	second (s)
sect.	section
sq.	subgenus
sp.	species
sp. gr.	specific gravity
T.S.	Transverse section
w/v	weight for volume, in the case of solids mixed with liquids.

INDEX

Abbreviations 133
Additional cultural characters 74, Plate 10.
Aegerita duthei 70
Agaricus spp.
    A. bitorquis 100
    A. brunnescens 98
    A. hortensis 49, 92-99, 101, Plate 13
Agrocybe spp. 46
    A. arvalis 73
    A. cylindraca 104
allocysts and related structures 78
    examples 79
Amanitopsis 62
anatomical features 52-54
antibiotics
      media 7
      supplements 13
arthroconidia 66-68
Antromycopsis 70
Armillaria 93, 107
    A. mellea 43, 44, 47, 74, 77, 79, 83, Plate 3 C.
asexual spores   see mitospores
Asterophora 69
Attamyces bromatificus 69

Biochemical tests 86-89
      instant tests 86-87
      long-term incubation tests 88-89
      short-term incubation tests 87-88
bioluminescence 79
Bolbitius 47
Boletus 51
    B. sulphureus 51, 69
branching pattern of hyphae 75-76
Brauniellula nancyae 43
bulbiloid bodies 72

Candelabra cells 78
caterva 73
    examples 80
cephalosus 73
    examples 80
check list of chemicals 132
check list of equipment 131
chemical tests 89

chlamydospores
        examples 71
        solitary 69
        terminal and intercalary 69
Chroogonphus rutilus 130
Clamp-connections 76-77
Collecting techniques 21-22
Collybia racemosa 70
    C. tuberosa 73
comparative studies 30
        chemical 74-75
        genetic studies 92-96
conidia 38, 41, 66-69, 95-96, Plates 8 & 9
conidial states 66
conidiophore 80
Conocybe 47, 68, 75, 92, Plate 3 B
contaminants (named) 43
Coprinus 39, 43, 47, 48, 68, 69, 73, 74, 75, 78, 95, 96, 101, 102, Plate 3 D
   C. cinereus 44, 47, 91, 96, 101-102
   examples of veil 110, Plate 11
coprophilous fungi 108-109
cultivation 42-43
cultural characters 59, 81
culture-formulae 81-83
cyphelloid fungi 63, Plate 15, Appendix 2 tables 1 & 2
Cystoderma 39, 68

Damp chamber 28, Plate 1 B
de-dicaryotization 58
delimiting pigments in hyphal walls 56
development of basidiome Plates 4-6
        angiocarpic 60
        bivelangiocarpic 61
        bulbangiocarpic 61
        examples 63-64
        gymnangiocarpic 61
        gymnocarpic 60
        hemiangiocarpic 60
        levhymenial 62
        metavelangiocarpic 61
        mixangiocarpic 61
        monovelangiocarpic 61
        range 59
        rupthymenial 62
        schizohymenial 61
        stipitangiocarpic 61
Digitellus 74

diverticulate 78

Ecological exercises 108-110
Entoloma abortivum 43

Field observations 21-22, Plate 1, Fig. C-E
fixatives: recipes 17-18
Flammulina velutipes 46, 47, 68, 83, 105,
fructification 46-51

Galerina mutabilis 104
gallic-acid test 85
gasteromycetoid fungi 62-63, Plate 14, Appendix 1 tables 1 & 2
genetic studies 90-96, Plate 12
gloeocystidia 78
glossary 116-122
Gomphidius roseus 43
Gymnopilus 68

Hebeloma 51
Hohenbuehelia 66
holoblastic conidia 66
       examples 72
homing response 95-96
Hypholoma 68

Inoculating 40-41
       basidiome 41
       basidiospores 41
       mitospores 41
       vegetative tissue 40
indirect isolation 39-40
       flotation method 39-40
       hyphal isolation 40
       soil dilution plates 39
       soil plates 39
isolation 31-40
       asexual spores 38
       basidiospores 35-38, Plate 2 D-J
       hymenial tissue 31
       indirect methods 39-40
       mycorrhiza 34-35
       pileus tissue 31, Plate 2 A
       soil 33
       stipe tissue 32-33
       vegetative phase 33-39
       wood-samples 34

Laccaria laccata 51
Lactarius rufus 51
Lepista nuda 44, 47
Lentinellus 69
Lentinula edodes 107
Lentinus spp. 33
    L. lepideus 74, 83
    L. tigrinus 83
Leucoagaricus 69
Lyophyllum 69
    L. ulmarium 83

Maceration 57-58, Plate 11 A-B
Maintenance of cultures 44-46
mating known/unknown strains 93-94
mating type determination 94-95
media preparation 8-9
    recipes 3-8
micro-stains
      nuclear stains 14
      recipes 14-16
mitospores 38, 41, 65-71, 95-96, Plate 8 & 9
modified clamp-connections 76-77
    examples 72
morphology of hyphae 75-77
Mycena spp. 68, 79, 92
   M. citriicolor 72, 79
Nematoctonus 66
Nothoclavulina 70
Nyctalis 69

Omphalotus olearius 79
Oudemansiella radicata 83
ozonium 74
    example 80
Ozonium auricomm 74

Paarige branching 76
Panaeolus 76
Panellus stipticus 79, 96
Panus badius 73
Paxillus involutus 51, 62, 73
Phaeolepiota 39
Pholiota spp. 47, 68, 69, 83
   P. nameko 104
physiological experiments 96-97
Pisolithus tinctorius 51

Pleurotus spp. 66, 70, 73, 77
    P.ostreatus 103, 106-107
Pluteus cervinus 35
Polyporus squamosus 74
preservation of voucher material 22-26
preservatives 20
Psathyrella 46, 48, 68, 92
pseudosclerotium 73
Psilocybe 46, 75
    P.cubensis 103
purification of cultures 42-44

recipes 3-8, 10-20
    antibiotic media 7
    buffer solutions 12-13
    chemically defined media 6-7
    commercial media 8
    fixatives 17-18
    micro-stains 14-16
    natural media 3-4
    non-nutrient media 7-8
    salt solutions 11-12
    slide mountants 16-17
    solid media 10-11
references 123-130
refractive bodies 78
restrictions to culturing 42-44
Rhacophyllus lilacinus 73
Rhizomorpha 74
rhizomorphs and similar structures 74
        examples 79
Rhizopogon parasitica 43

Schizophyllum commune 46, 78, 85, 91, 107-108
Scleroderma citrinum 51
Scleroma 51
sclerotia and related structures 73
        examples 79-80
Sclerostilbum 70
secondary spores see mitospores
single spore isolation 56-57
slide-culture 27-28, Plate 1 A
        mountants 16-17
        preparation 28-29
        preservation 29-30

Soil dilution plates 39

soil plates 39
solutions 11-14
    buffers 12-13
    saline 13-14
    salts 11-12
    trace elements 11
spot tests 89
Squamanita 70
stains
    nuclear 14
    recipes 14-15
    schedules 17-20, 51-54
staining
    dolipores 56
    nuclei 55
    siderophilic granules 56
    tissues 51-54
sterile fruitings 74
    examples 81
Stilbum flavidum 73
Stropharia 68
S. ferrii 102
Suillus spp. 43, 51, 62, 76

Tannic-acid test   see gallic acid
techniques
    field 21-26
    genetical 56-58
    laboratory 26-51
    pure culture 30-40
    routine 26-30
    staining 51-54
thallic arthroconidia 66-68
    examples 70-71
tissue types 77

Volvariella esculenta 104-105

Xeromphalina campanella 83

PLATE I. FIG. A. Cover-slip culture: cover-slip inserted under agar and slits cut to permit removal of agar and cover-slip. Preferable to mark the under surface with grease pencil.

FIG. B. Damp chamber: a. greased Petri dish; b. cover-slip; c. glass slide; d. filter-paper; e. inoculum; f. agar block.

FIG. C. Taking spore-print by cutting off stipe, laying on a glass slide (a) and piece of clean paper and enclosing in a piece of waxed paper.

FIG. D. Taking spore-print by threading the stipe through a card held over a glass vessel.

FIG. E. Sealing plastic bags by sandwiching between strips of glass and melting overlapping plastic.

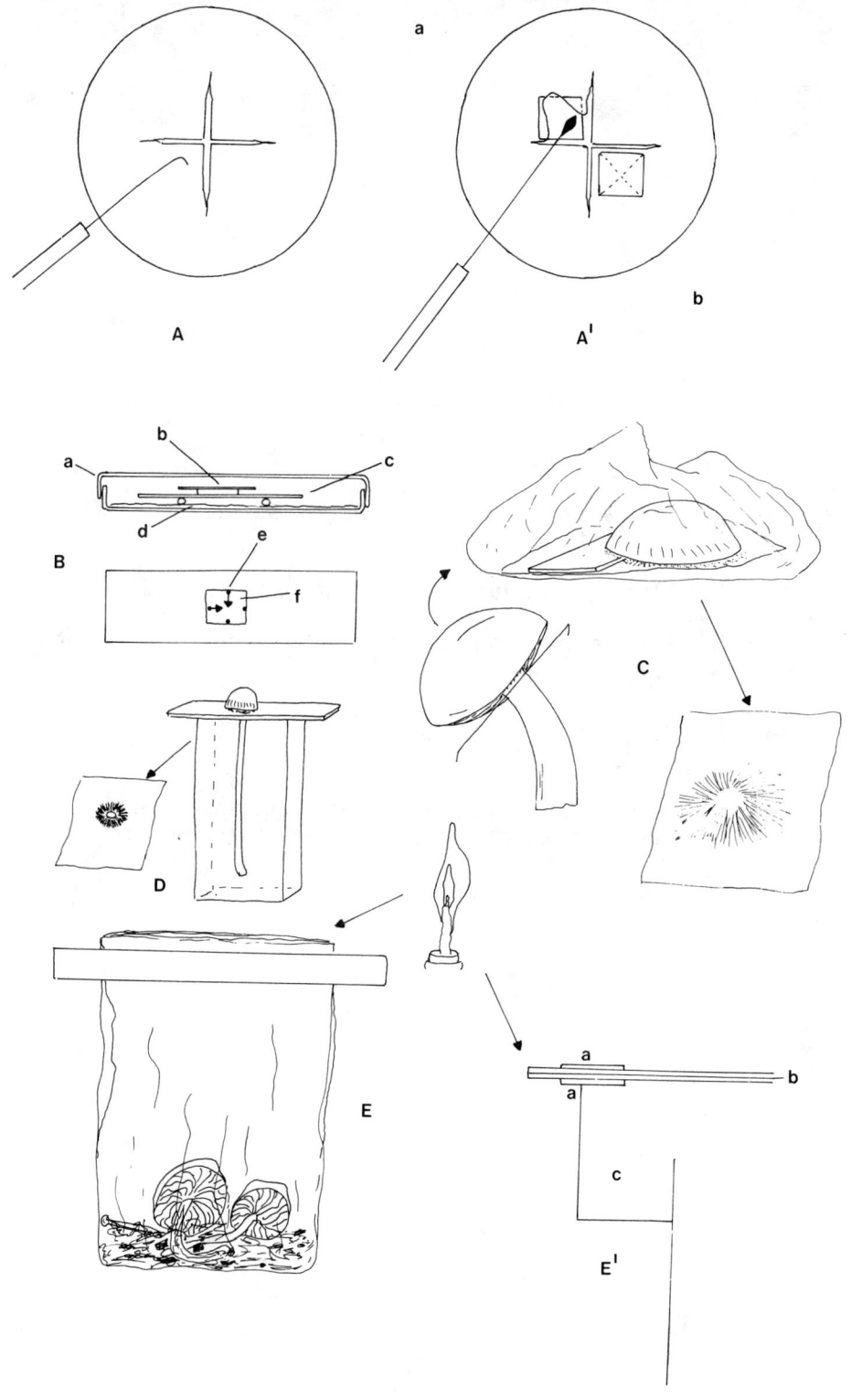

PLATE 2. FIG. A & B. Taking inoculum from halved basidiome, preferably split and not cut.

FIG. C. Placing inoculum on agar surface in Petri dish.

FIG. D. Cutting out from basidiome a section with gills attached. (D^1).

FIG. E. I. cutting agar slope, turning the slope tip and pushing it down to lie over the remainder of the slope.

FIG. F. Removal of small piece of agar and small section of basidiome 12 hours later. Spore-print on agar surface.

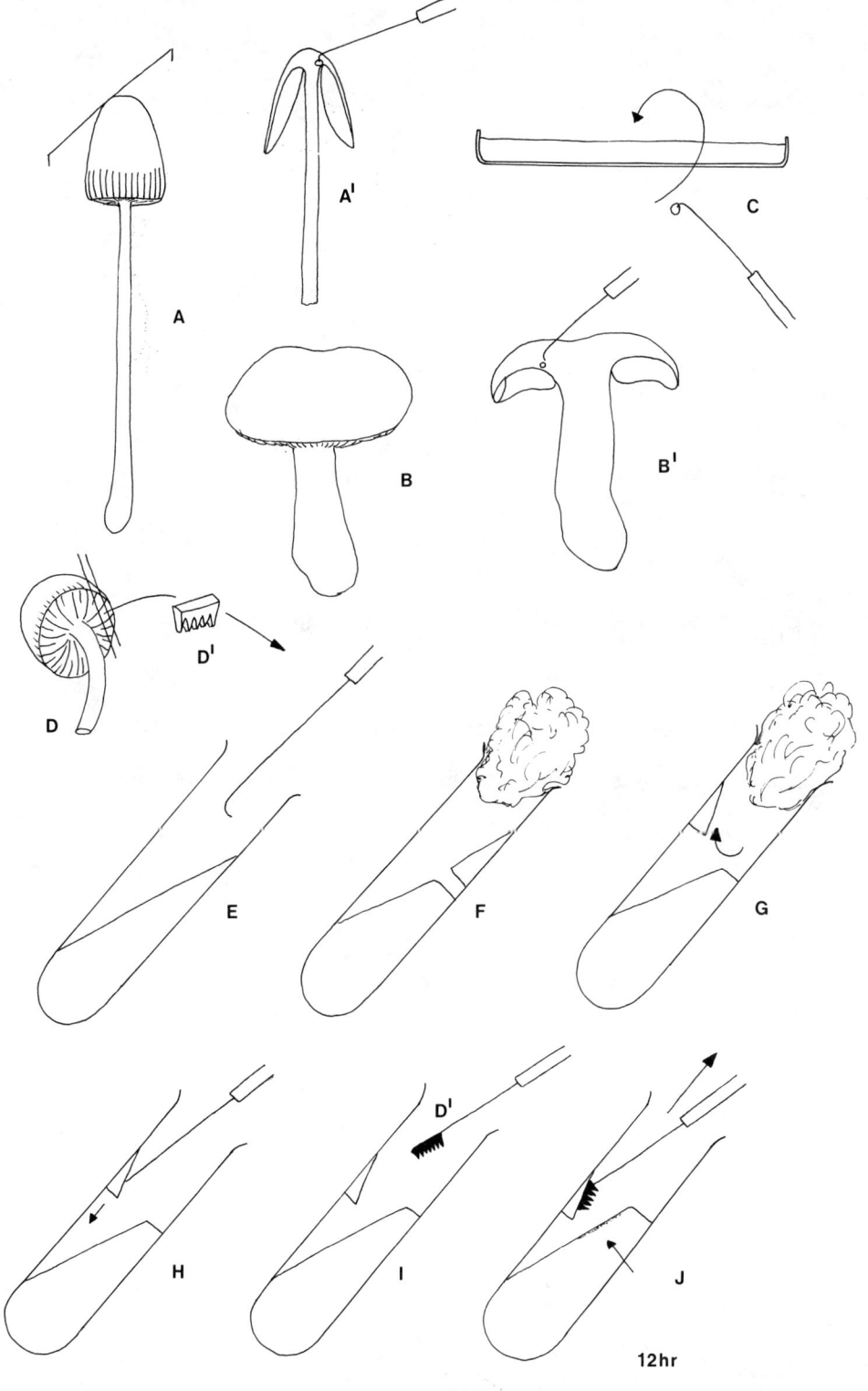

PLATE 3. Fructification. Photographs by R. Eudall, Royal Botanic Garden, Edinburgh.

FIG. A. *Pholiota terrestris* on paper pulp with added 2% malt extract agar.

FIG. B. *Concybe farinacea* on sterile dung.

FIG. C. *Armillaria mellea* on sterile corn (maize) kernels.

FIG. D. *Coprinus* aff. *extinctorius* on paper pulp mixture.

A, B & D subject of Ph.D. study; originals in R.B.G. collection, Edinburgh.

A

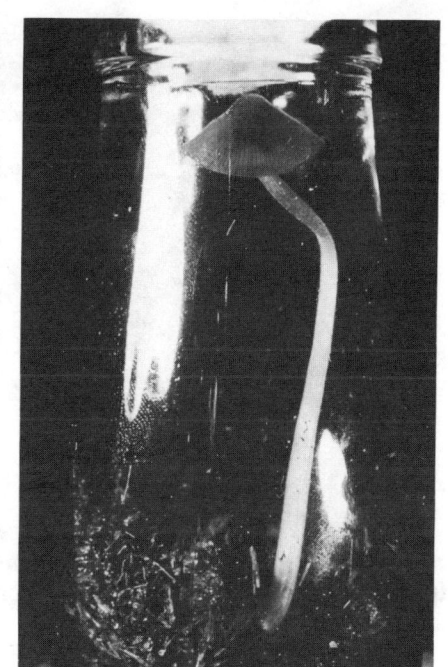

B

C

D

PLATE 4. Gross-developmental types and stages of development of basidiome of *Amanita phalloides*, with variation in volvate bases found in *Amanita* spp. A. & B. immature basidiome bursting through volva (section:C); D. ring beginning to break; E. mature basidiome; F. base of *Amanita citrina;* G. base of *Amanita muscaria;* H. base of *Amanita fulva.*

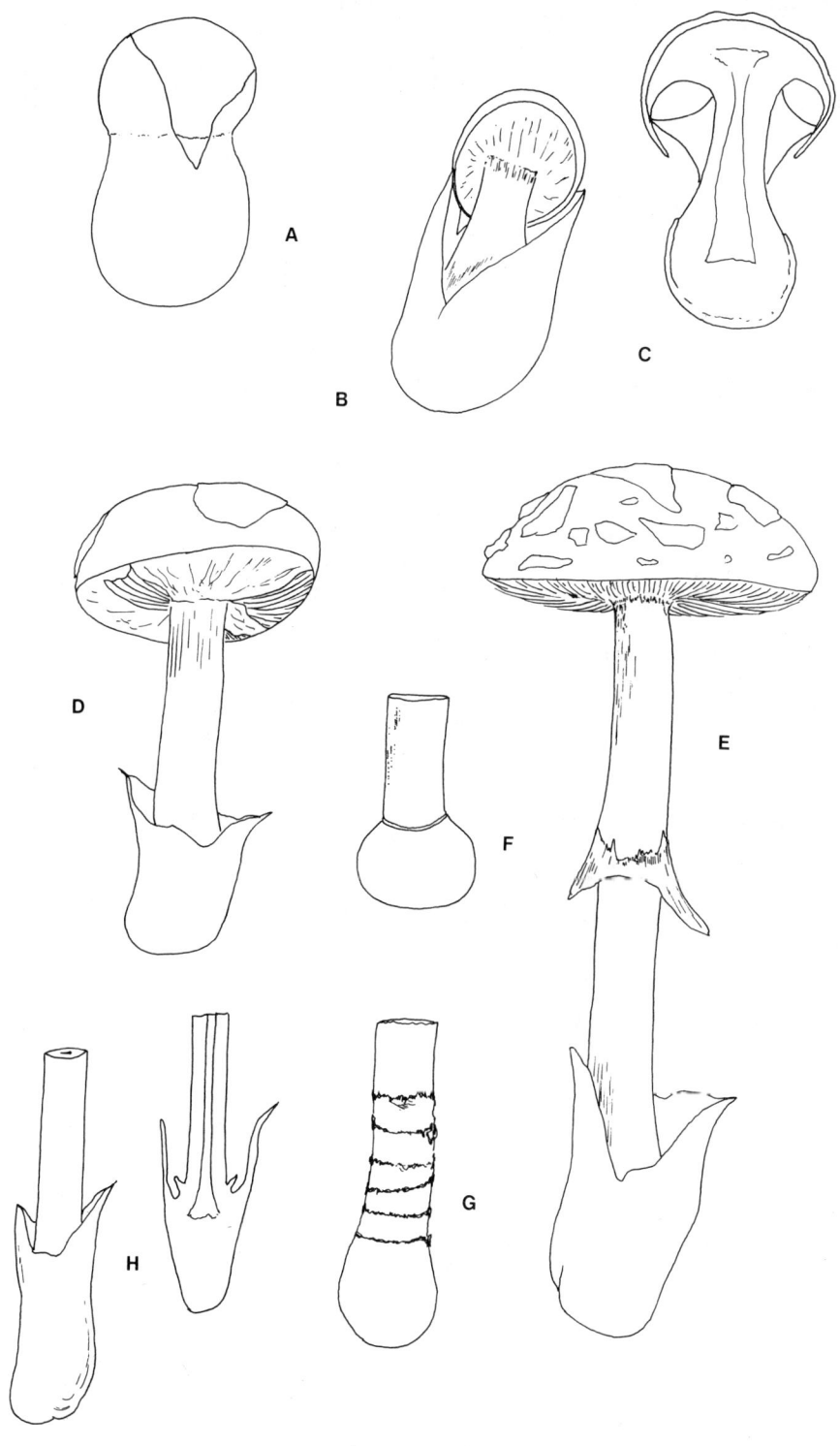

PLATE 5. FIGS. A-E. Gymnocarpic and related developmental categories recognised by Reijnders.

FIGS. V-Z; development from small primordium.

FIG. A. strictly gymnocarpic.

FIG. B. mixangiocarpic—further growth of pileipellis.

FIG. C. mixangiocarpic further growth of stipe.

FIG. D. gymnocarpic with down curling of pileus.

FIG. E. mixangiocarpic, further development of both pileipellis and stipe.

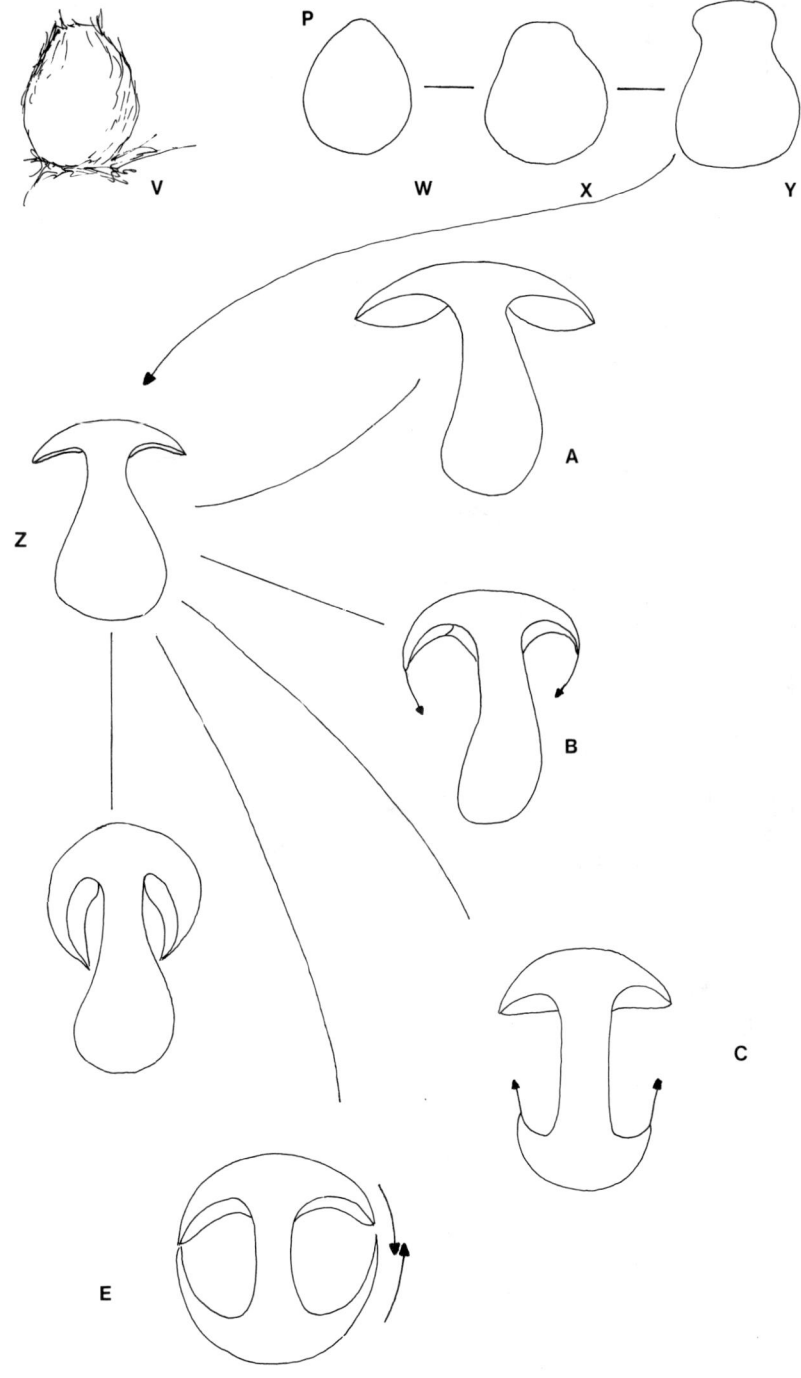

PLATE 6. Hemiangiocarpic (Pseudo-velangiocarpic) developmental categories recognised by Reijnders. Development from small primordium (X).

    FIG. A. metavelangiocarpic.

    FIG. B. bulbangiocarpic.

    FIG. C. stipitoangiocarpic.

    FIG. D. pilangiocarpic.

    FIG. E. hemiangiocarpic.

    FIG. F. endocarpic (=angiocarpic).

    FIG. G. gymnocarpic.

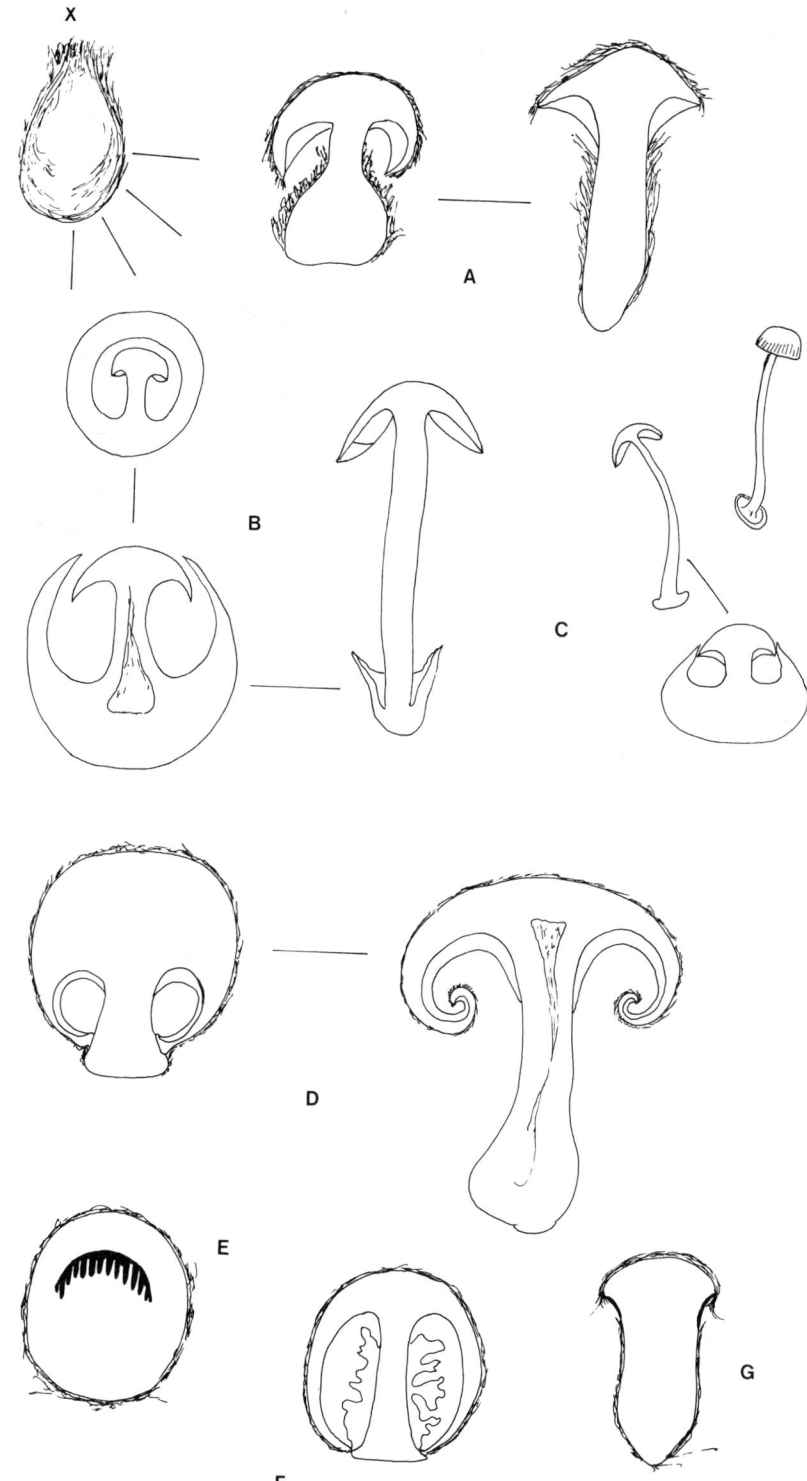

PLATE 7. Hemiangiocarpic (velangiocarpic) developmental categories recognised by Reijnders. Development from small primordium (X) —monovelangiocarpic (A); paravelangiocarpic (B); gymnangiocarpic (C); bivelangiocarpic (schizohymenial) (D); bivelangiocarpic (levhymenial) (E). Both D & E lead to F.

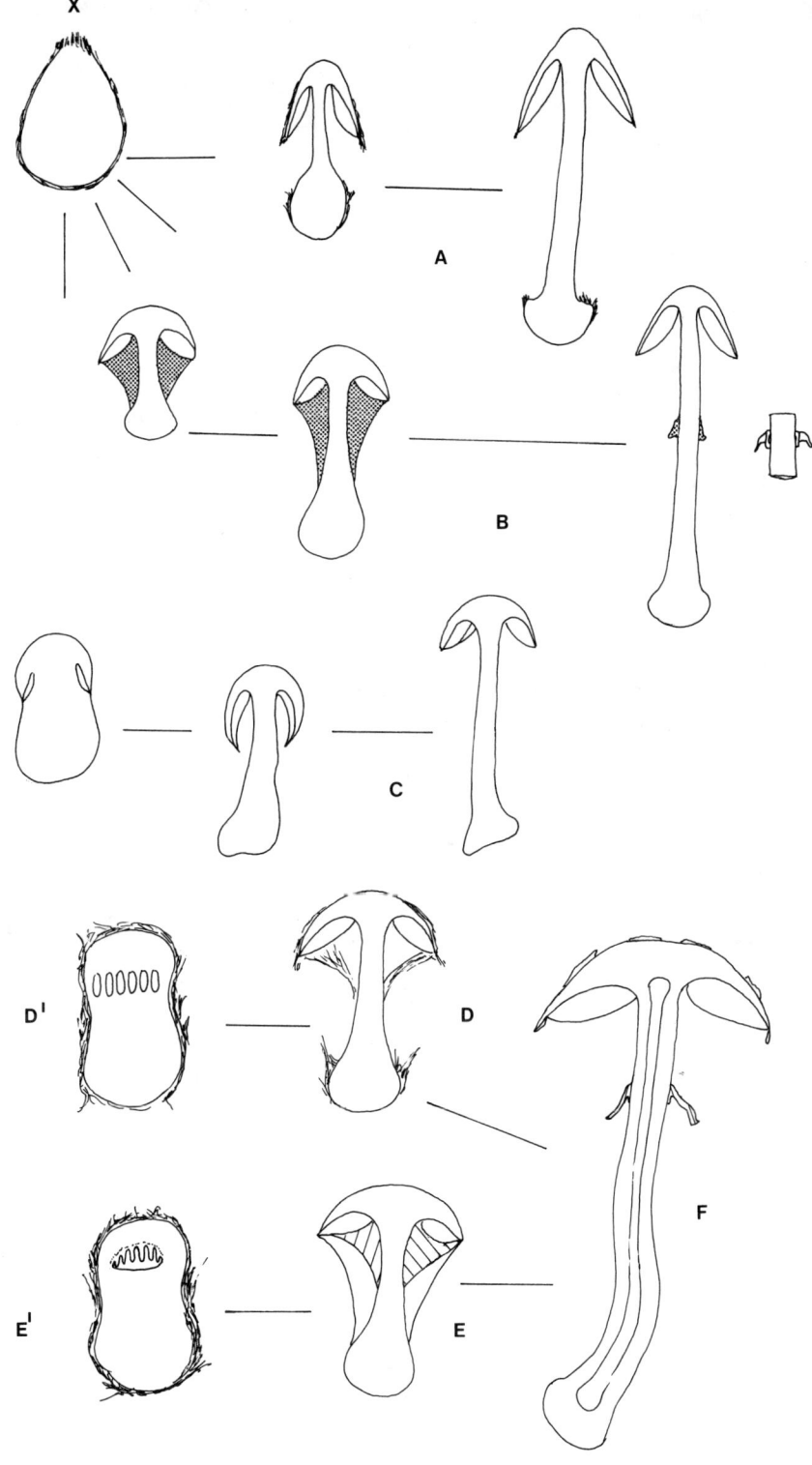

PLATE 8. Mitospores — 1.

    FIGS. A—E thallic arthroconidia.

    FIG. G. *Nyctalis asterophora (Asterophora lycoperdoides)*.

    FIG. H. *Nyctalis parasitica (Asterophora parasitica)*.

    FIG. C. Chain from Tricholomataceae *(Mycena inclinata)*.

    FIG. D. Chain from Strophariaceae *(Psilocybe merdaria)*.

    FIG. E. Curls from Bolbitiaceae *(Agrocybe semiorbicularis)*.

    FIG. F. Chain of chlamydospores in Tricholomataceae *(Lyophyllum decastes)*.

    FIG. I. Chain with branches in Bolbitiaceae *(Conocybe farinacea)*.

    FIG. J. Sympodial conidium in *Squamanita odorata*.

    FIG. K. Sympodial thick-walled conidia in *Dissoderma paradoxa*.

    FIG. L. Sympodial thick-walled, ornamented conidia in *Squamanita pearsonii*.

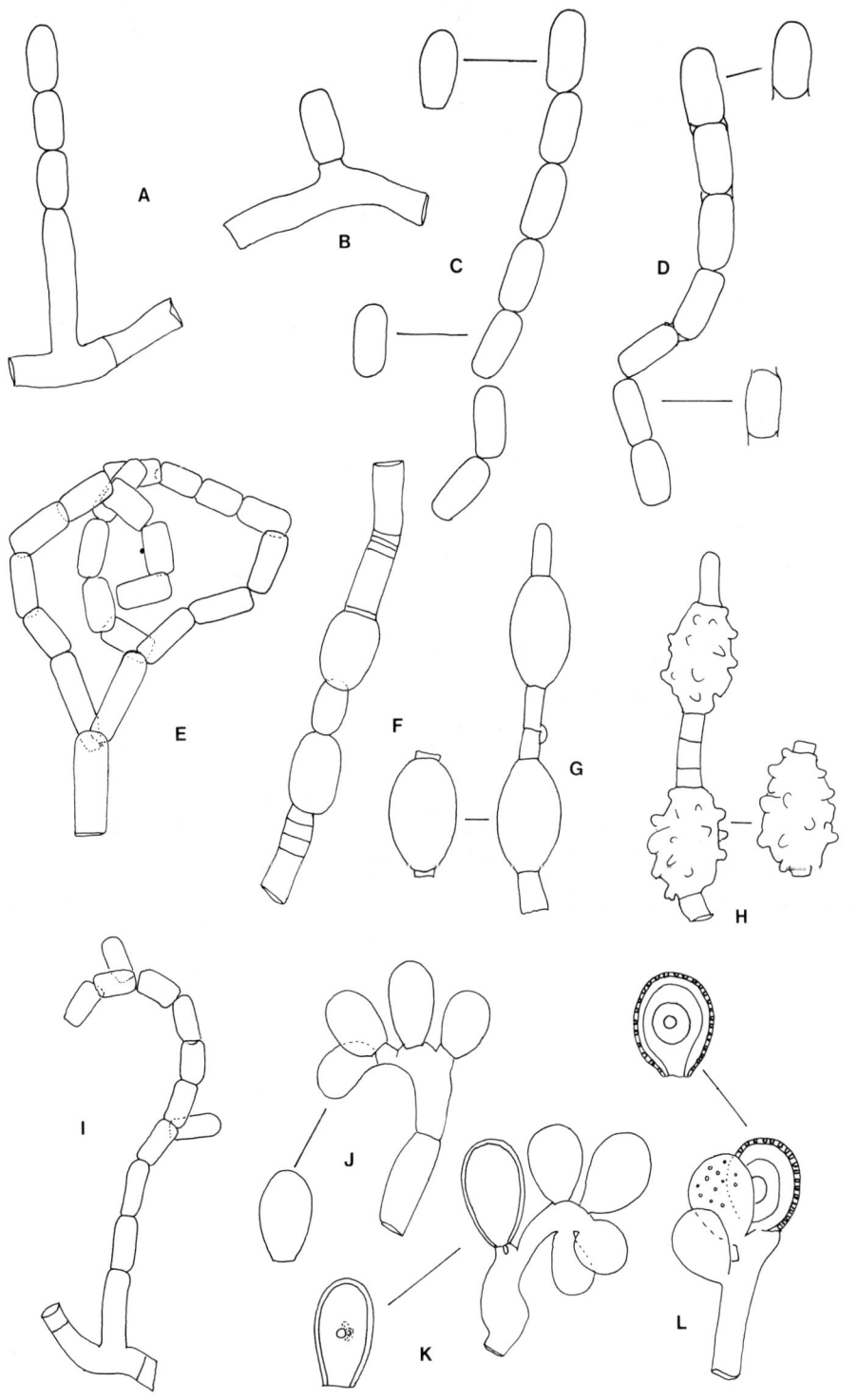

PLATE 9. Mitospores — 2.

    FIG. A. *Attamyces bromatificus*.

    FIG. B. *Pholiota aurivella*.

    FIG. C. *Hohenbuehelia longipes*.

    FIG. D. *Collybia racemosa* with stipe coremia magnified (E) (*Sclerostilbum septentrionale*).

    FIG. F. *Arthrosporella ditopa* with *Nothoclavulina* (G).

    FIG. I. *Pleurotus cystidiosus* with *Antromycopsis broussonetiae* (H).

    FIGS. J & K. *Rhacophyllus lilacinus* with catervae (L & M) and its teleomorph *Coprinus clastophyllus* with section showing basidia.

    FIG. P. *Mycena citriicolor* with *'Stilbum' flavidum* (Q)—a dispersing cephalosus (R).

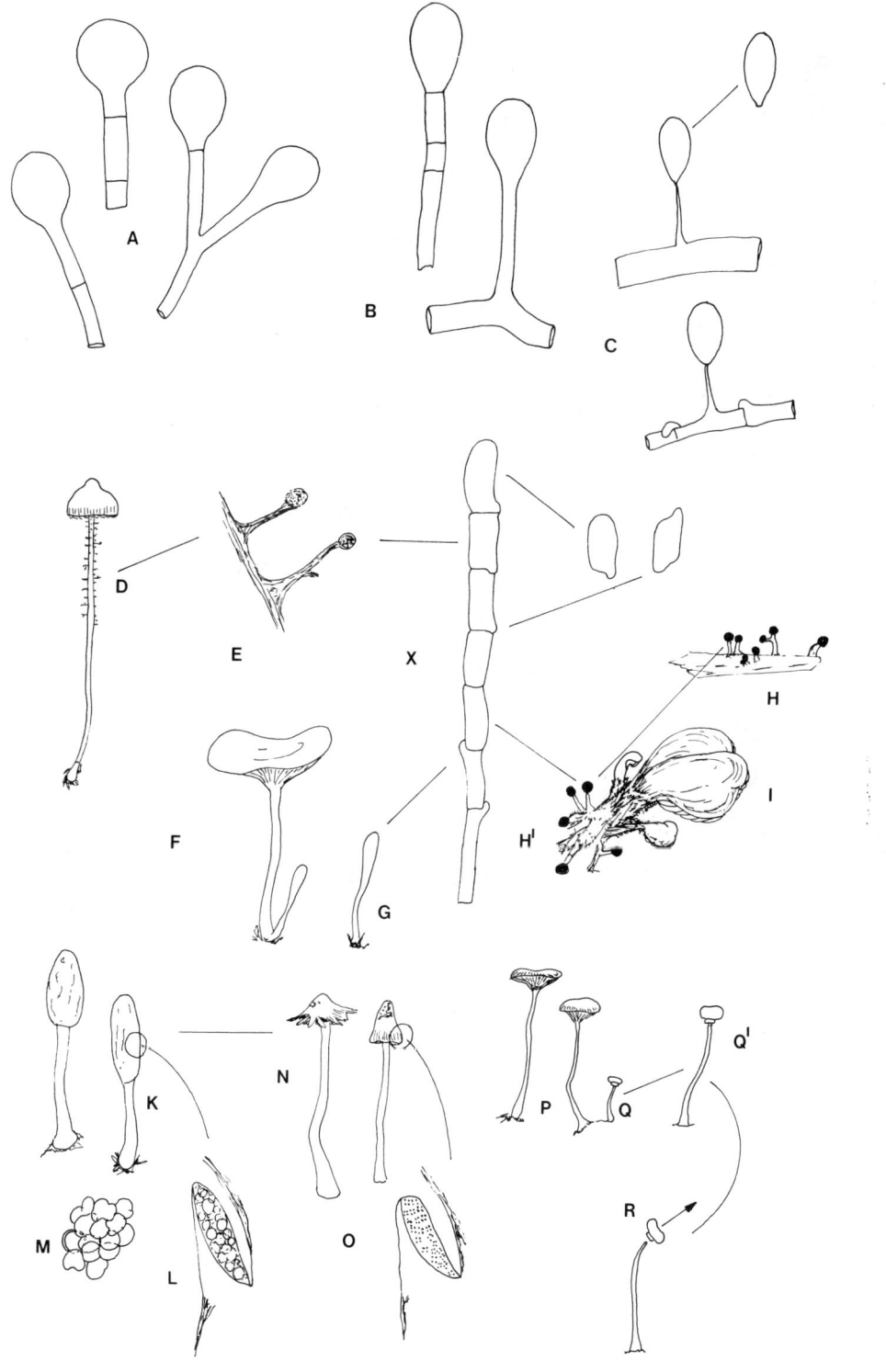

PLATE 10. Additional cultural characters.

    FIG. A. Thick-walled hyphae with simple septa.
    FIG. B. Stephanocysts.
    FIG. C. 'Bulbils': knots of hyphae.
    FIG. D. *Schizophyllum commune:* small diverticulae with apical droplets.
    FIG. E. Diverticula with crystalline apex.
    FIG. F. Chain of swollen thin-walled cells.
    FIG. G. Plate-like crystals on hyphae.
    FIG. H. Amorphous granular material on hyphae.
    FIG. I. Needle-shaped crystals on hyphae.
    FIG. J. Nodose-septate hyphae (= clamp-connected hyphae).
    FIG. K. Gloeohyphal vessels.
    FIG. L. Setules: Fig. M. Seta.
    FIG. N. Paarige branching in *Suillus*.
    FIG. O. Thick-walled diverticulate hyphae.
    FIG. P. Laticiferous hyphae.
    FIG. Q. Thick-walled skeletal-like hyphae in *Pleurotus dryinus*.
    FIG. R. Allocysts.
    FIG. S. Thick-walled pseudoparenchymatic cells.
    FIG. T. Diverticulae on rounded cells of *Coprinus stercoreus*.
    FIG. U. Pseudoparenchymatic cells.
    FIG. V. Generative hyphae showing dolipores accentuated by congo red stain in ammoniacal solution; note pseudo-clamp.

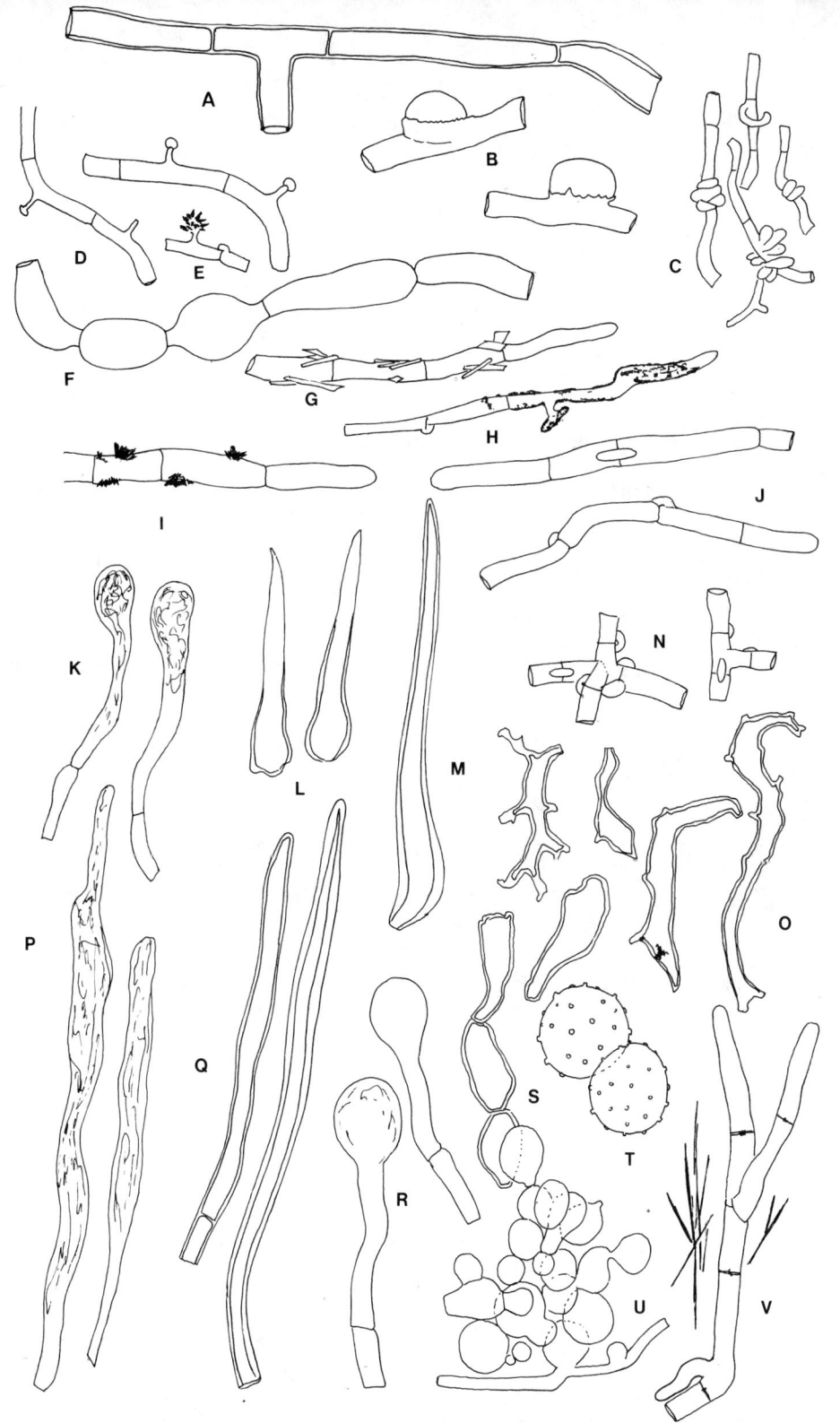

PLATE 11. Maceration techniques and velar types in *Coprinus*.

FIG. A. Glass tissue grinder with small amount of sterile water or saline solution in bottom.

FIG. B. Plunger of grinder with small piece of culture.

FIG. C. Culture plate indicating source of inoculum.

FIG. D. Action of plunger to macerate tissue followed by shaking (D^1) to form suspension.

FIG. E. Pouring out of hyphal fragments on agar surface.

FIG. F. Cells of veil of *Coprinus: Lanatuli*.

FIG. G. Cells of veil of *Coprinus: Impexi*.

FIG. H. Cells of veil of *Coprinus: Vestiti* with acid-resistant ornamentation (diverticulate).

FIG. I. Cells of veil of *Coprinus: Vestiti* with acid-soluble ornamentation.

FIG. J. Cells of veil of *Coprinus: Micacei*.

FIG. K. Pileus of *Coprinus: Setulosi*.

FIG. L. Pileus of *Coprinus: Hemerobii*.

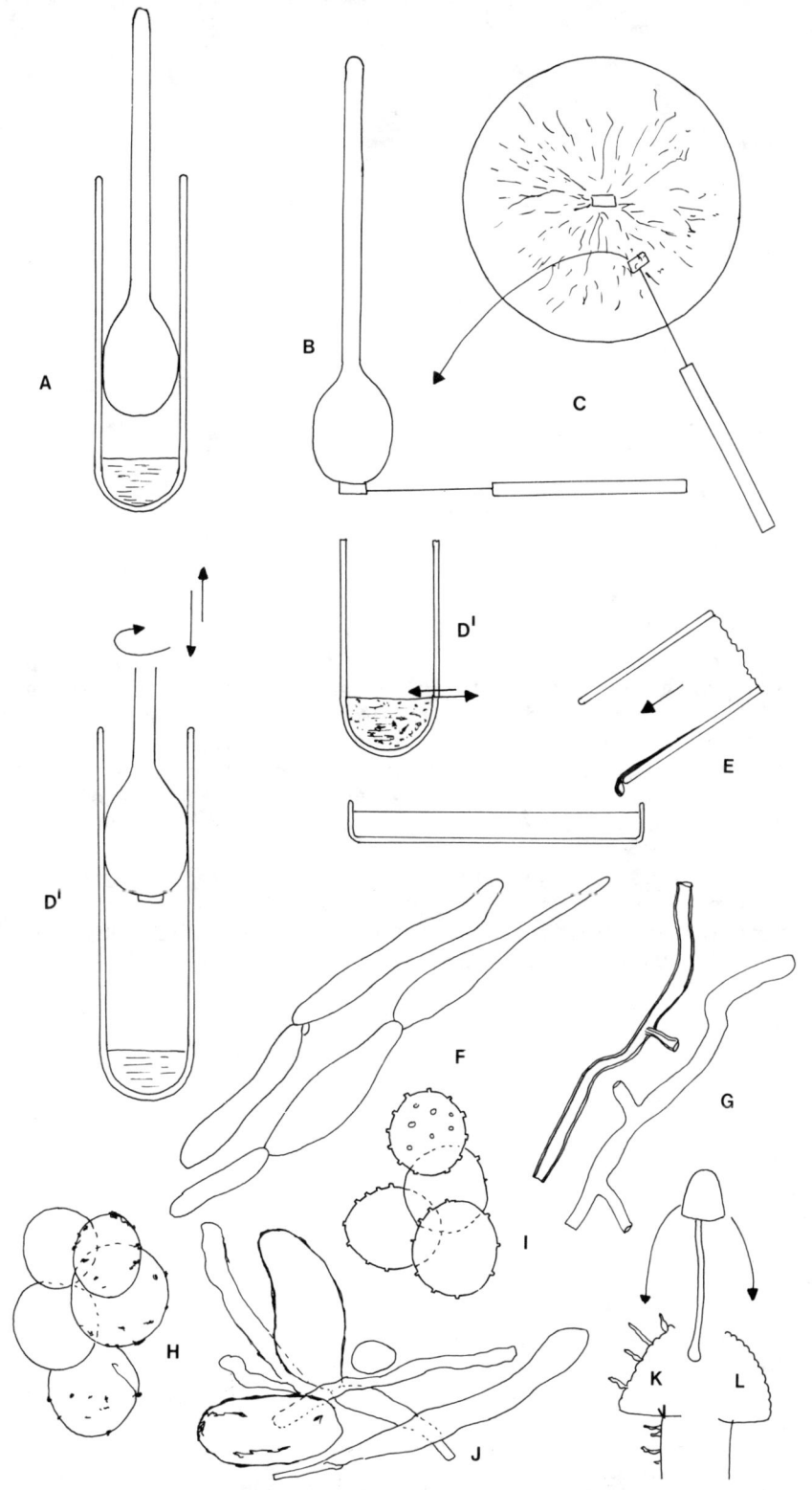

PLATE 12. Schematic life-history of a typical heterothallic agaric. Agaric (t) producing cloud of basidiospores (s) from the basidia (b) produced by the gills (g) under the pileus. Two spores A & B either produce a primary mycelium forming thallic arthrospores (c) or fuse in pairs to form a secondary mycelium exhibiting clamp-connections, and producing either uninucleate thallic arthrospores (d) or binucleate conidia (e). Ultimately this mycelium produces basidiomes, in which the fusion nucleus forms.

The lower figures show the way in which oidia may fuse with hyphae of the opposite mating type and in one case exhibit a lethal reaction with vacuolation of the apical cells. A homothallic agaric can complete the cycle in Plate 12 with only one basidiospore, either A or B.

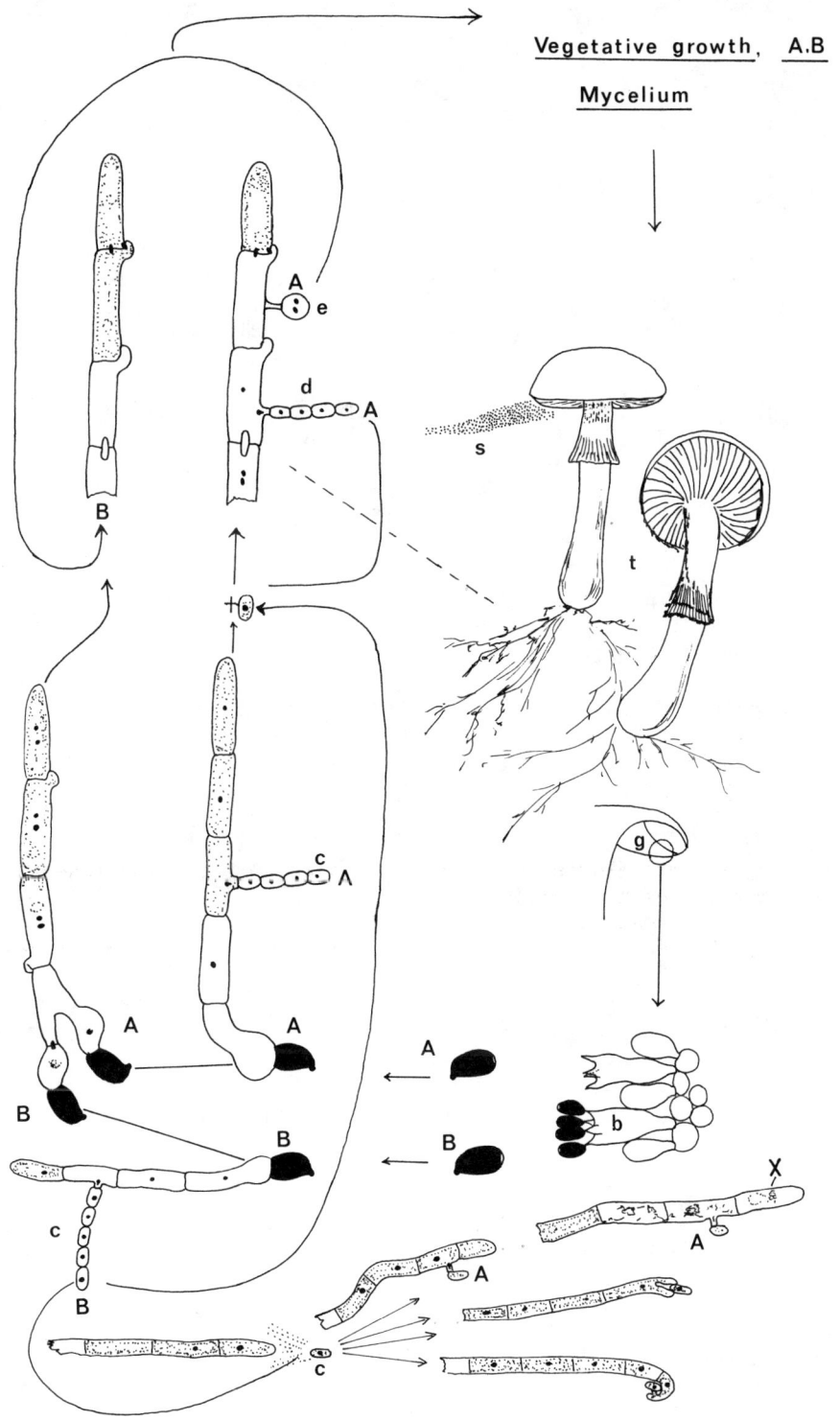

PLATE 13. Experiments with *Agaricus hortensis*.

FIG. A. Development of the basidiome of *A. hortensis*, showing the areas of greatest elongation and expansion. The black dots represent similar dots which can be placed on mushroom buttons to monitor the development amongst a home-grown product.

FIG. B. Section of cultivated mushroom (a) seated on a pad of damp blotting paper (b) in a suitable container (c).

FIG. C. A slice of basidiome indicating the area from which to take the stimulating tissue (d).

FIG. D. Parallel slice to C. but with gills on one side removed. The resultant curvature is recorded in Fig. E.

FIG. E. Resultant curvature.

FIG. F. Slice with gills on both sides removed and a portion of gill (d) placed at the trimmed end of one side.

FIG. G. Resultant curvature.

FIG. H. Slice with a piece of agar to which has been added an alcoholic extract of gills prepared by vigorously shaking up some gills in ethanol for 10 minutes.

FIG. I. Curvature; water agar is placed at the opposite end.

FIG. J. Slice as in H with water agar placed at both ends and to one side a piece of gill-tissue attached as in D - E.

FIG. K. A piece of gill-tissue attached to the stipe of a **developing** basidiome with a piece of agar.

FIG. L. Curvature produced.

FIG. M. A whole basidiome is cut and and inert substance placed in the cut, e.g. a mice flake (e).

FIG. N. Shows the complex expansion produced.

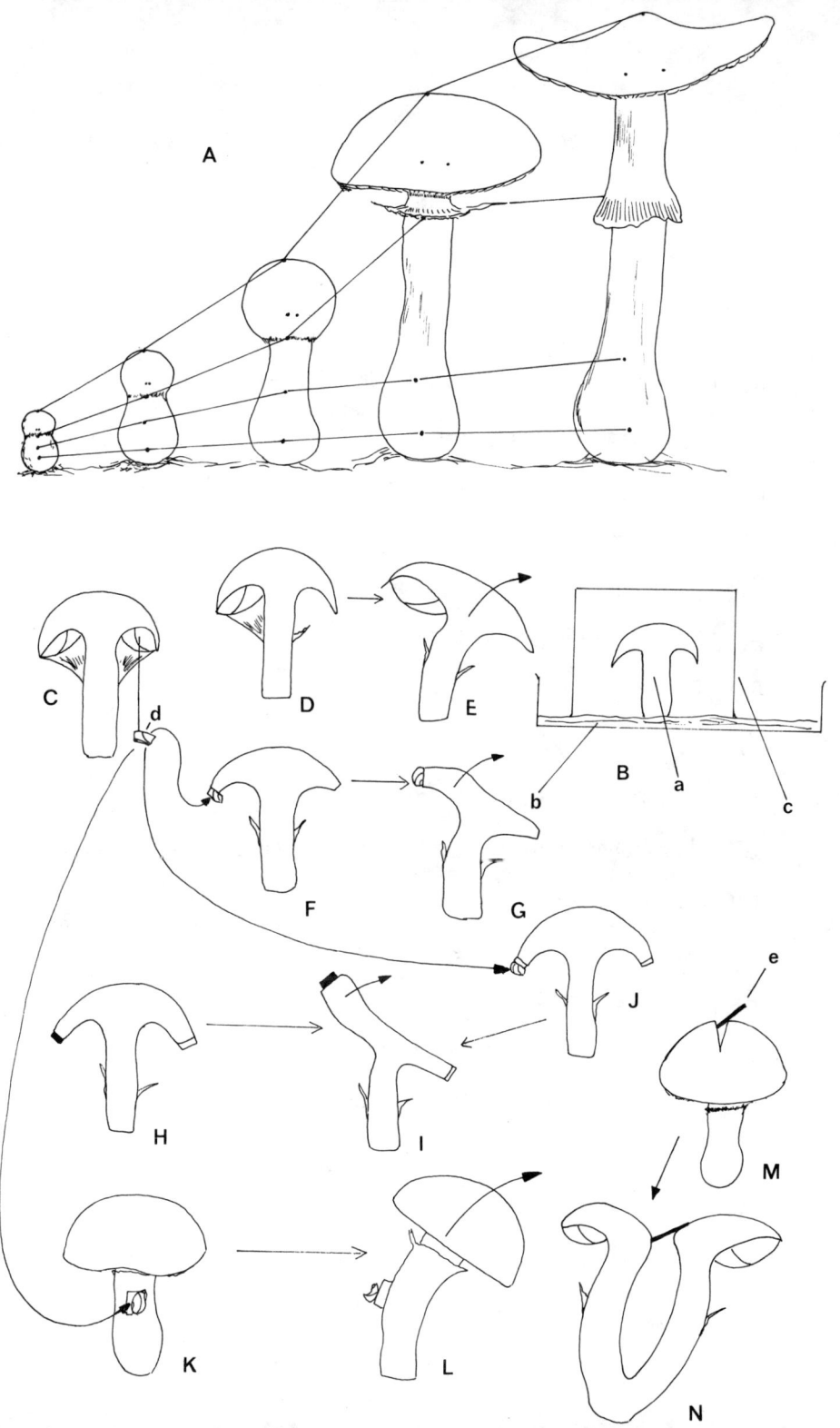

PLATE 14. Gasteroid agarics and boletes; Basidiomes and sections.
    FIG. A. *Weraroa coprophila.*
    FIG. B. *Galeropsis polytrichoides.*
    FIG. C. *Brauniella.*
    FIG. D. *Paxillogaster.*
    FIG. E. *Cortinarius kashmiriensis.*
    FIG. F. *Martellia.*
    FIG. G. *Elasmomyces.*
    FIG. H. *Torrendia pulchella.*
    FIG. I. *Richonella.*
    FIG. J. *Rhodogaster.*
    FIG. K. *Thaxterogaster pingue.*
    FIG. L. *Macowanites.*
    FIG. M. *Gastroboletus scabrosus.*
    FIG. N. *Gastroboletus turbinatus.*

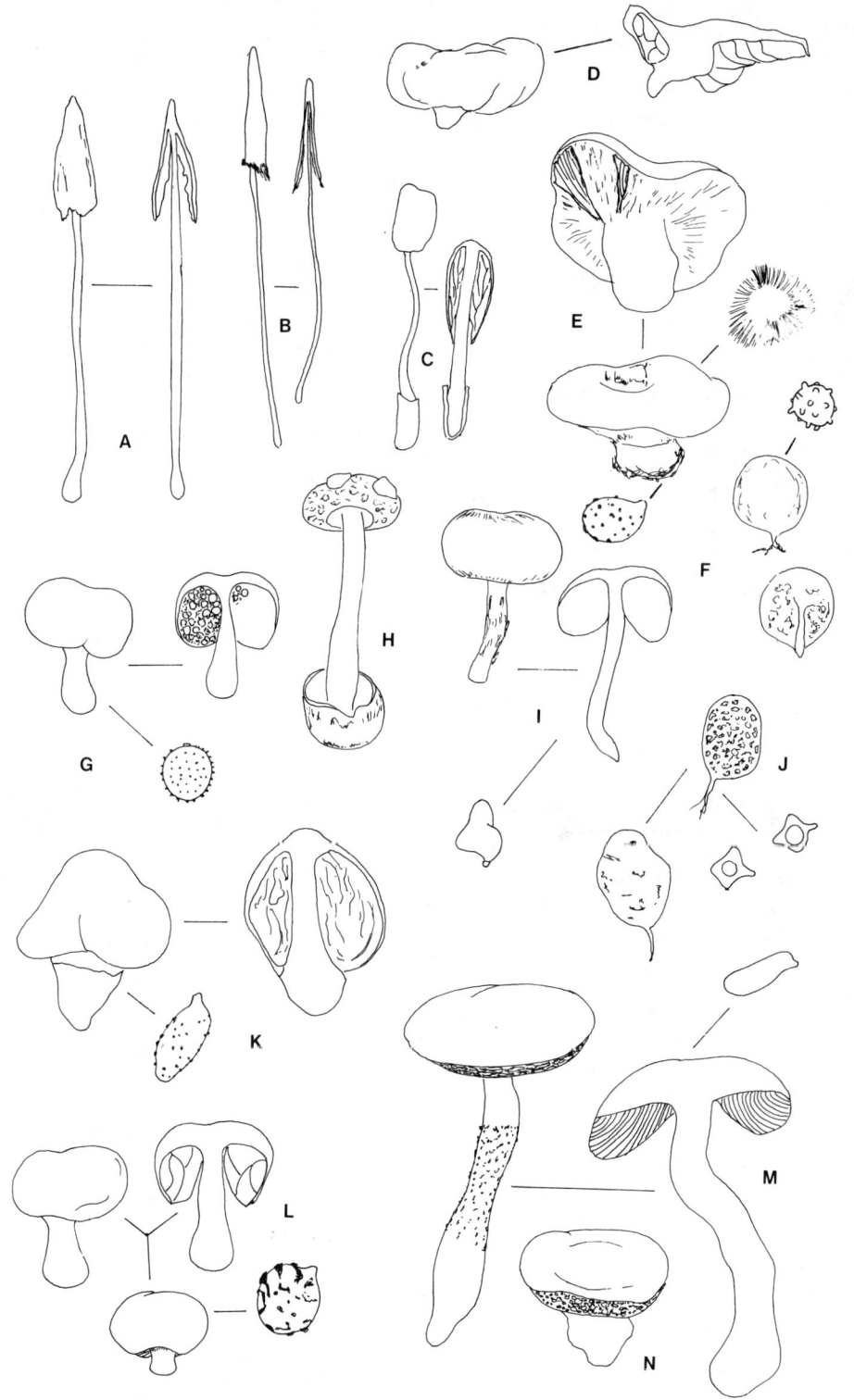

PLATE 15. Cyphelloid agarics

    FIG. A & B. *Flagelloscypha* (B^1 hairs from basidiome).

    FIG. C. *Cyphellopsis confusa* (C^1 hairs from basidiome).

    FIG. D. *Henningsomyces pubera* (D^1 hairs from basidiomes).

    FIG. E. Hairs of *Cellypha goldbachii*.

    FIG. F. *Gloiocephala menieri*.

    FIG. G. *G. caricis*.

    FIG. H. *Skepperiella*.

    FIG. I. *Arrhenia auriscalpium*.

    FIG. J. *Favolaschia*.

    FIG. K. *Chromocyphella muscicola*.

    FIG. L. *Cyphellopsis anomala*.

    FIG. M. *Phaeotellus acerosus*.

    FIG. N. *Cyphellostereum muscigenum*.

    FIG. O. *Panellus: P. pusillus* (P) and *P. stipticus* (P^1).

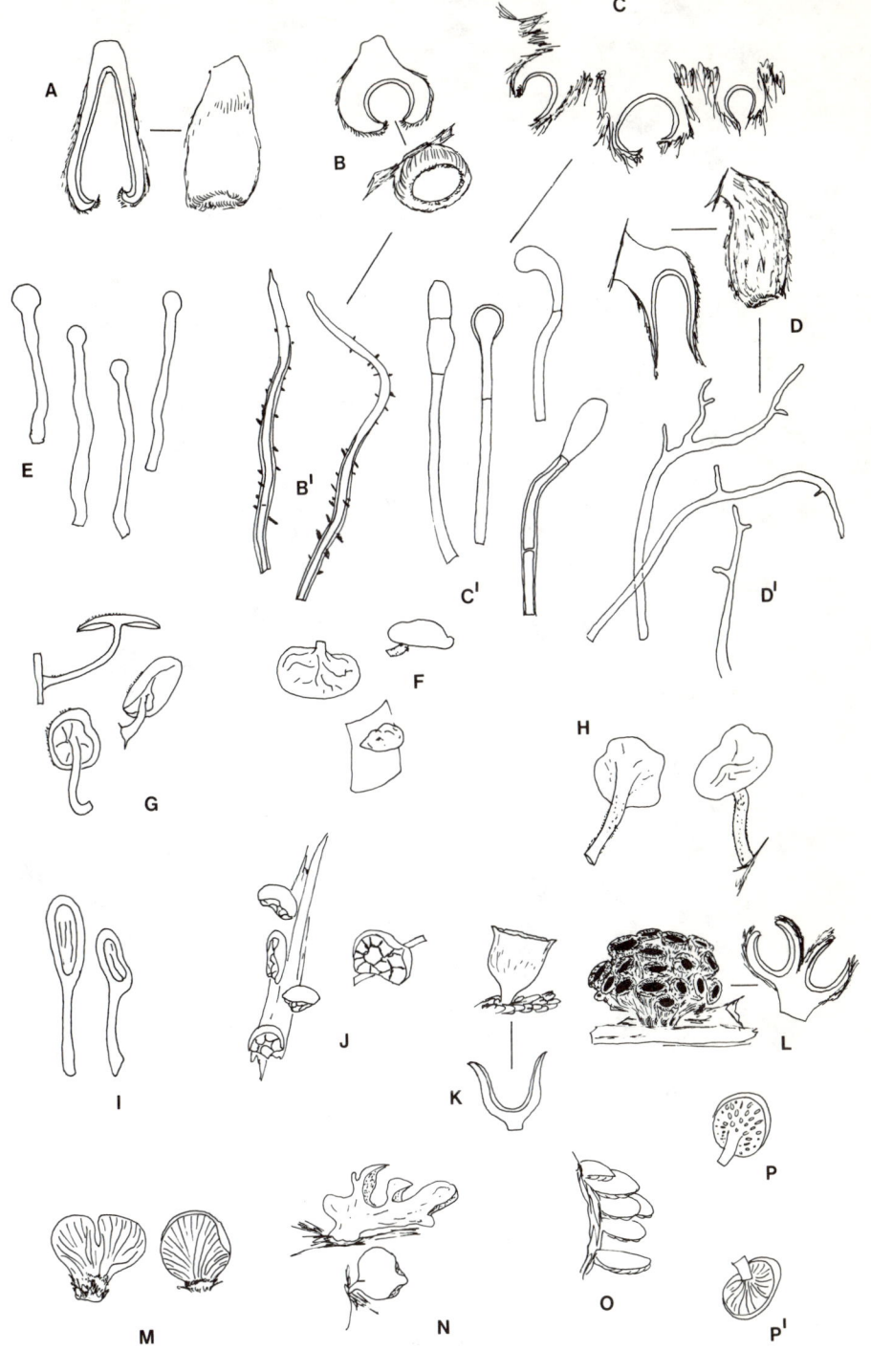